PROPHECY AND POWER

Beyond the mountains there are more mountains;
  although they appear to be disconnected, actually they are not.
Beyond the trees there are more trees;
  although they appear to be connected, actually they are not.

Chinese aphorism

World is crazier and more of it than we think,
Incorrigibly plural.

Louis MacNeice

Can there be anyone in the world who has not got mixed
  feelings?
Should there be anyone in the world who has not got mixed
  feelings?

Paul Durcan

PROPHECY AND POWER

ASTROLOGY IN EARLY MODERN ENGLAND

Patrick Curry

Princeton University Press
Princeton, New Jersey

Published by Princeton University Press
41 William Street, Princeton, New Jersey 08540

Library of Congress Cataloging-in-Publication Data

Curry, Patrick
    Prophecy and power: astrology in early modern England / Patrick
Curry.
      p.   cm.
    Bibliography: p.
    Includes index.
    ISBN 0–691–05579–3
    1. Astrology—England—History—17th century. 2. Astrology-
–England—History—18th century. 3. England—Intellectual
life—17th century. 4. England—Intellectual life—18th century.
I. Title.
BF1679.C87 1989
133.5′0942′09032—dc20   89–33447
                                                                   CIP

Printed in Great Britain

Contents

Abbreviations

Add. MS	Additional Manuscript
Ash. MS	Ashmolean Manuscript
Bacon	J. Spedding, R. L. Ellis, and D. D. Heath (Eds), *The Works of Francis Bacon* (London, 1857−59; 14 vols)
BDBR	R. L. Greaves and R. Zaller (Eds), *Biographical Dictionary of British Radicals* (Brighton: Harvester Press, 1982−4; 3 vols)
BL	British Library
Bodl.	Bodleian Library
Bowden	Mary Ellen Bowden, 'The Scientific Revolution in Astrology: The English Reformers, 1558−1686' (Yale University, unpublished Ph.D. thesis, 1974)
Capp	Bernard Capp, *Astrology and the Popular Press: English Almanacs 1600−1800* (London: Faber, 1979)
CSPDS	*Calendar of State Papers, Domestic Series* (London)
DNB	Sir Leslie Stephen and Sir Sidney Lee (Eds), *Dictionary of National Biography* (Oxford: Oxford University Press, 1921−2; 22 vols)
DSB	C. C. Gillispie (Ed.), *The Dictionary of Scientific Biography* (New York: Charles Scribner's Sons, 1974; 16 vols)
ESTC	*Eighteenth Century Short Title Catalogue*
Josten	C. H. Josten (Ed.), *Elias Ashmole (1617−92). His Autobiographical and Historical Notes, His Correspondence, and other Contemporary Sources relating to his Life and Work* (Oxford: Oxford University Press, 1966; 5 vols)
Phil. Trans.	*Philosophical Transactions of the Royal Society*
RS, FRS	Royal Society, Fellow of the Royal Society

Taylor   E. G. R. Taylor, *The Mathematical Practitioners of Tudor and Stuart England* (Cambridge: Cambridge University Press, 1954)

Thomas   Keith Thomas, *Religion and the Decline of Magic* (Harmondsworth: Penguin, 1973; first published 1971)

(References in the Notes supply name, date, volume (if any) and pages. Fuller information about the source is given in the Bibliography. Almanacs are given by the year covered, not the year published.)

Acknowledgements

It is a real pleasure to thank those people who have helped me with this book – beginning with my parents, for their big-hearted support over many years. In addition, I am grateful to four people in particular: Suzanna, my wife and best critic, who has shared much of the practical brunt of its writing; Simon Schaffer, for reliable guidance and enlightenment on a frighteningly wide range of subjects; Angus Clark, for his moral support and literary acumen; and Piyo Rattansi, for skilfully and patiently steering me through my Ph.D. on this subject.

For very helpful discussions and suggestions, I would like to thank the following: Ernesto Laclau, Geoffrey Cornelius, Michael Hunter, Ellic Howe, Ann Geneva, Roy Porter, Jacques Halbronn, Olgierd Lewandowski, Stuart Clark, Clay Ramsay, Garey Mills and Ray Keenoy.

I have also received much help from the staff at the many libraries, especially the British and Bodleian; from the generosity of the Social Sciences and Humanities Research Council of Canada, who funded a final six months of research on the eighteenth century; and, not least, from Pip Hurd at Polity Press.

Finally, my thanks to The Blackstaff Press (Belfast), publishers of Paul Durcan, and Faber and Faber (London), publishers of Louis MacNeice, for their kind permission to quote from these two poets' works.

1

Introduction

This Book

Astrology has long aroused strong feelings. In a millenia-long history, it has attracted as many as it has repelled, springing up afresh after every apparent defeat whether by science, religion, or simply 'reason'. Astrologers have even made bold to claim those attributes for their own. Is it then unrecognized science, or divinatory magic? Prophecy or prediction, or simply placebo? The very impossibility of supplying a single or definitive answer has lent astrology much of its longevity, and its interest. Such adaptability and elusiveness is thus not simply a problem; it is what makes astrology worth studying.

I became interested at an early age. Since then, my attitude has changed radically several times: from partisan, to sceptic, and finally to historian. Like Ellic Howe, I eventually realized that 'the answers to the questions that interested me most – these questions had nothing to do with validity or otherwise of astrological beliefs – would not be discovered by casting horoscopes, but by a process of historical research.'[1] In this process, I found the positions of both partisan and sceptic equally unhelpful; not only do I lack the necessary hubris to declare the transcendent truth about astrology, but also I believe that neither attitude can begin to do justice to astrology's rich and subtle complexity. For that, one must learn to remain (in the words of an aptly placed literary critic) 'faithful to the ambiguity of opposing demands . . . [refusing] any single place or position which would permit the illusion of a final solution.'[2]

After all, what we call astrology has itself changed many times, and radically so, since its Babylonian and Greek origins. To put it another way, astrology has never been just one thing. But the period and place discussed in this book – from 1642 to 1800, in England – offer particular problems and rewards. As the astrologer William Lilly remarked in 1666, 'The English of all nations are most taken with prophecyes.'[3] Throughout the last six centuries, English men and women have considered the meaning of the stars, cast horoscopes, and consulted astrologers. In the late seventeenth century, however, a dramatic transformation took place. Astrology fell from unprecedented influence, during the English Revolution, to what is often described as its death, after the Restoration. Yet this phenomenon is seldom discussed by historians. The obvious exception is Keith Thomas's pioneering *Religion and the Decline of Magic* (1971) – for very good reasons, still an indispensable introduction to the subject. A mark of its fruitfulness, however, is that Thomas's book raised as many questions as it answered. One purpose of this book is therefore to take them on, and suggest answers: did astrology really die, or decline; in so far as it did, why and how did that happen? What happened afterwards? And what can we learn here?[4]

Taking such questions seriously – and their human subjects – is also part of a larger project: that is, to recover astrology from what E. P. Thompson memorably described as 'the enormous condescension of posterity.'[5] He was referring to posterity's neglect of the labouring poor and powerless, but it seems to me that the point is the same: to contest the notion of history as fundamentally about the 'winners', by making room for those now deemed to be its 'losers'. In any democratic view of history, this is just as important whether we are looking at people's beliefs and world-views, or their ways of life; they cannot really be separated.

I have tried to bring a forgotten cast – the astrologers (of whatever kind) and their clients, as well as the critics – to life. They deserve no less, and our common history would be poorer without them. An equally vital part of history-writing, however, is to hazard risky (but fruitful) explanations. Only in this way can we arrive at a deeper understanding of the past, and, by implication, the present.[6] In the case of astrology in early modern England, the potential rewards extend well beyond a better grasp of the 'precise sociological genealogy' of its decline.[7] For example, this history throws new light on the interactions between patrician, middling, and plebeian cultures – especially the dynamics of hegemony, whereby the ideas and values of the first class dominated (to various degrees) those of the other two. Again, seeing astrology in its historical context brings out the intimate relationships between highly specific social and political events, on the one hand, and enduring mental attitudes or 'mentalities', on the other. Paradoxically, a more historical approach to such issues can sharpen our thinking about them in general.

It is not really surprising that the history of astrology should offer such rewards. The stars were widely considered to be God's handiwork, and therefore a source of divine guidance; so astrology overlapped with religion. As part of natural knowledge, their study fell within the ambit of natural philosophy, medicine, and early modern science. Astrological predictions frequently had political repercussions. In all these ways – as with any system of knowledge and belief (that is, epistemology) which is rooted in the lives of real people, living in a material world – astrology was inseparable from considerations of power.[8] It can therefore act as a sensitive historical instrument, its rejected and forgotten 'truth' throwing new light on that of the dominant, successful and mainstream.[9] But, as I learned myself, in struggling to find the right approach, such insights are lost if astrology is seen as simply one of history's 'losers'. Although speaking *de haut en bas*, such an historian is reduced to demonologist and hagiographer, chronicling the ignorant believers and prescient critics of a delusion. This approach is what I regard as teleological – the idea that history has a goal towards which it is marching, aided by heroes and obstructed by villains – and anachronistic: the willingness to re-enter history as a vicarious participant, rewarding and punishing with hindsight.[10]

For reasons that lie partly in the very period under study here, such an attitude (in more or less sophisticated variants) is a particular problem in the study of the occult, magic, or 'pseudo-science'. As elsewhere where it appears (usually among certain adherents of the Marxist left and New Right), the primary purpose is to use history in order to advance points of contemporary ideology. With astrology and the like, that usually means scientific ideology; and its commonest marker is an uncritical use of the word 'superstition' and its synonyms. There is no reason why the historian should try to deny contemporary political or social interests; they are unavoidable, even desirable, and better explicit than suppressed. But they do not justify bad scholarship. And that is unavoidable, if the historian insists on taking up an exclusive position inside the very historical debate that we are trying to understand. Conceived as popular religion, astrology was first stigmatized as superstition (mystification, ignorance, and so on) by the Reformation churches; most recently, this label has been adopted and adapted by scientific apologists. But as James Obelkevitch put it,

'Before popular religion can be understood at all, it has to be understood in its own terms, not as a failed version of something else.' The same principle applies with equal force to astrology considered as a so-called 'pseudo-science.'[11]

Another purpose of this book is therefore to offer an alternative example of the social history of such ideas, based on a critical pluralism. In both subject-matter and approach, a premise is that truth, rationality and progress are never self-evident. Rather they are themselves the valued prizes contested (usually with distinctly unequal resources) among historical subjects: astrologers, divines, scientists, politicians, plebeians, patricians alike. None had, or have, privileged access to their possession; or, to put it another way, their truth-claims must be treated identically. As G. E. R. Lloyd stated: 'The explanandum is not, in any case, the victory of rationality over magic: there was no such victory: but rather how the criticism of magic got some purchase.'[12]

Partly to this end, I have adopted a working definition of astrology to include any practice or belief that centred on interpreting the human or terrestrial meaning of the stars. This definition is sufficiently flexible to allow the full phenomenon to flower, since astrology was no more a single entity with a transcendently fixed character than was, or is, 'science' or 'religion'.[13] Any move to make it appear so substantively, beyond semantic convenience, is therefore a rhetorical intervention in a particular debate, whether one in history or now.

Astrology does have a central feature, however, which such a definition brings out. That is the act of interpretation – in this case, of the stars: from astral plant and weather lore of the simplest and most popular kind, or the divinatory revelations of judicial astrologers, to the cosmic speculations of eminent natural philosophers. Such interpretation did not take place in a vacuum; as we shall see, it was intimately connected to other considerations of knowledge, values and power. It is not, therefore, an example of autonomous or free-floating cultural 'discourse'.[14] On the other hand, astrological interpretations unavoidably embodied ambiguity and contingency. For that reason, they were always open to re-interpretation and re-articulation by others, for different ends. Thus, despite their connectedness, astrological interpretations were never fixed, or uniquely and finally determined (whether by economic or other supposedly isolated factors). Personally, I see this kind of openness – which applies equally to historical and other acts of interpretation – not so much as a problem as a sign of life and hope.[15]

4

Astrology in Historical Context

I have space for only the briefest sketch of the economic, social and political setting in early modern England. Throughout this period, England was a predominantly rural and pre-industrial society. In the mid-seventeenth century, the total population was four to five million, of which about 80 per cent (according to the pioneering demographer Gregory King) lived in villages or in the countryside. The exception was London. But even with a population of just under half a million, in 1665 – at least twenty times greater than any of its British urban rivals – London could claim only about eight out of every hundred people in the country. Nor was the capital immune from rural afflictions; plague swept through it in 1603, 1625, and, most disastrously, in 1665.

The social world in this period was finely and hierarchically structured, with huge differences in conditions and styles of life between those at the top and the bottom of the social ladder. Although there were no mature social classes of the kind we associate with modern industrial society, it would be highly misleading to suggest that England was a classless or so-called 'one-class' society. Accounts by contemporaries such as William Harrison and Gregory King suggest the following broad social groups: at the top, the aristocracy and landed gentry (a very small number, with a great deal of status, wealth and power); next, the professional and mercantile class, composed of clergy, lawyers, physicians, and the wealthier merchants, and largely based in the towns; and, at roughly the same 'level', the rural parish gentry; then, rural freeholders, the better-off tenant farmers, yeomen, and the tradesmen, shopkeepers, and artisans; after them, labourers, servants, poor husbandmen, and common soldiers (a very large class, with very little); and finally, an underclass of paupers, beggars, and vagrants. King estimated that the landowning and professional elite received a larger share of the national income than all the other groups added together.[16]

Here too, London was unusual, without radically breaking the overall pattern. From the late seventeenth century onwards, it was the setting for a dramatic rise in the number of people of 'middling' status and wealth, especially among the professions and trades. Illiteracy too was much lower in the capital than elsewhere. With its concentration of political power and cultural influence, London played a disproportionate role in those aspects of the country as a whole.[17] Nonetheless, English society in general was still more local and varied than can easily be imagined today.

The second half of the seventeenth century in England saw extraordinary events. They involved and eventually transformed the entire country, through probably the most traumatic thing that can befall one (short of subjugation): a civil war. That war broke out in 1642, and lasted until 1649, when the reigning monarch, Charles I, was beheaded – a deeply shocking event, even for many of his opponents, and one which reverberated through the following century of English history. Almost as disturbing (or exhilarating) was the displacement of Anglicanism as the national Church. Following rule by Parliament, then the army, then Cromwell (as Lord Protector), monarchy, Church and Lords were restored in 1660.

The significance of the Restoration, against the backdrop of the preceding two decades, is likewise difficult to overestimate. On the one hand, it was a real restoration of royal and clerical power; on the other, that power could never again be as 'natural' and impregnable as it had been before the war. The intervening changes ruled out any simple return to the previous period, and the subsequent strife-torn rule of the Stuarts ended in the Glorious Revolution of 1688–89, bringing in its train a new settlement that cemented the marriage of monarchy, a national Church, and middle-class mercantile muscle. As we shall see, these momentous events and processes affected astrology no less strongly than everything else in the life of the country. Nor can other developments be ignored, such as the appearance of institutionalized natural philosophy in the the Royal Society (itself deeply implicated in the politics of the age).

Events in the eighteenth century can be seen in large part as a continuing response to those of the seventeenth. The consolidation of agrarian capitalism (including enclosures) accelerated the historic division between labour and capital, and the growth of a new entrepreneurial commercialism. The social analogue of these changes was the emergence and strengthening of a new gulf between patrician (upper and new middle classes) and plebeian (labouring class). Increasingly, the metropolitan literati became an effective voice of the former, and, through the rapidly growing metropolitan and provincial press, another instrument of London's political and cultural dominance. Politically, the brief Tory rule under Anne was followed by the long and increasingly conservative Whig and Low Church ascendancy after the accession of the Georges in

1714. By this time, the Anglican Church was deeply divided between High and Low, and mainstream politics into Whig and Tory. In mid-century, the Church united in the face of new competition from popular religion, especially Methodism; and the two parties, too, moved closer together after the passing of Jacobitism as a serious threat. Popular disturbances in the 1760s failed seriously to shake the ruling elite, but late in the century, Jacobinism and popular millenarianism raised new fears of sectarian radicalism – never very far from the surface, in any case.

Existing accounts of English astrology in this period agree that the mid-seventeenth century was a high-water mark. Astrologers and their almanacs flourished as never before. Beginning with the Restoration, however, an abrupt and serious decline apparently set in. In his discussion of the reasons, Thomas cites intellectual changes brought about by new scientific discoveries and technological advances, urbanization, and finally a new 'ideology of self-help'. At the same time, however, he admits the weakness of these explanations,[18] which indeed becomes apparent under close scrutiny: many well-informed astronomers and mathematicians were also astrologers; astrology adapted eagerly to newsprint and increased literacy, in the city no less than the countryside; and it was a personal weapon of advancement in the armoury of more than one self-made man of the times. Nor does religious criticism alone explain any decline; clerical hostility was more or less a constant throughout the entire period. Bernard Capp's later work underlines the problem. He emphasizes (correctly) that astrology lived on among the labouring classes of the eighteenth-century, but admits that 'we can point to no new or compelling argument or discovery to explain why astrology lost its hold over the educated classes.'[19]

I think this particular mystery can now be laid to rest. Astrology did see a rapid and striking change between 1660 and 1700 – although nothing that could, in good conscience, still be called a death[20] – and in what follows, we shall first concentrate on the late seventeenth century in order to bring out the dynamics of this transition. Then we shall look at the world of eighteenth-century astrology as a (reconstructed) whole. As will emerge, the latter period simply cannot be understood without the former. This approach will involve enlarging the relatively narrow definition of astrology used by most historians – mostly what I call judicial, which was the site of the more dramatic changes – in order to recognize and appreciate its popular and elite uses. Two caveats: first, in order to present and explore a neglected subject, I risk giving the impression of going to the other extreme, and implying that astrology was everywhere; clearly it was not; second, this study is intended to map the terrain and leave some reliable signposts. Along the way, I hope it will emerge that the history of astrology in early modern England is richly rewarding, and far from entirely inexplicable. But there is still much to explore, particularly in relation to the eighteenth century.

Astrological Theory and Terminology

My goal here is to provide the reader with a sufficient minimum grasp of astrology for this book. Much fuller guides are available.[21] Even the theoretical part of astrology's history is millenia-long and highly complex; what follows is therefore a bare but, I hope, useful outline.

It was a medieval and early modern commonplace that, broadly speaking, there were two kinds of astrology: natural and judicial.[22] The former concerned astral influences on 'natural' phenomena, such as the weather and agriculture, and (an indication of intellectual assumptions rather different to our own) human events of a mass nature, from epidemics to sweeping political and religious changes. Supposedly, only general and probable (as opposed to supposedly certain) predictions were undertaken. The close relationship of this astrology to natural philosophy or early science – with its emphasis on causal and law-like determinism – is obvious, and has been stressed by historians such as Thorndike and Neugebauer.[23]

Judicial astrology, on the other hand, embraced relatively precise astrally derived predictions or advice, concerned with specific individuals. ('Predictions' and 'prophecies' were, in this context, effectively inter-changeable terms; whether one or the other was used depended on the intentions of the astrologer or critic, but no clear distinction was sustainable.) The branches of judicial astrology divided into the following: nativities (also called 'genethliacal' astrology), based on a map of the planets' positions at the moment of a person's birth; horary, which endeavoured to answer questions (usually practical, often urgent) based on a map of the planets' positions at the moment when the question was asked, or received; and elections and inceptions, which concerned the beginning of an enterprise. In all cases the map was called a 'figure' – in the case of maps for persons, 'nativities' or 'genitures' – and later, 'horoscopes'. Considered as a practice, judicial astrology was further removed than natural astrology from natural philosophy, and was more a pragmatically oriented craft. (Bearing in mind that then, as now, any philosophy had practical implications, and, conversely, that practices were inseparable from philosophies, in this case cosmology in particular.)

The natural–judicial polarity, however, crossed with another important distinction: natural vs supernatural. This consideration derived largely from the unstable mixture of Aristotelian and neo-Platonic ideas that astrologers had inherited from the Renaissance and antiquity. The former contributed the order of the planets, descending from the primum mobile and the fixed stars through Saturn, Jupiter, Mars, the Sun, Venus, Mercury, and the Moon, down to the sublunary Earth. ('Planets' were commonly understood to include the Sun and Moon, and 'stars' was a generic term for both stars and planets.) The Earth was thus the maximum recipient of celestial influence, which acted through the four elements (fire, earth, air and water) and their corresponding Galenic humours in the human body (choler, black bile, blood and phlegm).

Whereas this contribution to astrological discourse tended towards naturalism, the more supernatural neo-Platonic strain appeared in concepts of sympathy and antipathy. These were vital to astrological magic and, arguably, divination. They were also used in physic, indicating (for example) that an illness with Saturnian symptoms (cold, stiffness, and so on) should be treated by herbs or medicines ruled by the opposite planet, the Sun, which therefore possess warming, relaxing virtues; and so on. The same ideas were also influential in attaching numerological significance to certain numbers in astrological interpretation, particularly the aspects (see below).

A strain-line thus ran among astrologers, roughly dividing adherents of Aristotelian naturalism from those of neo-Platonic magic. (This situation was further complicated by the rise of the new experimental philosophy in the late seventeenth century.) And judicial astrology, more than natural, tended to be seen as operating more by supernatural, or at least extra-natural, means. (This assertion was often hotly, if predictably, contested by its practitioners.) Even within judicial astrology, horary was held to be closer to the supernatural end of the scale. The inability of astrologers to produce a convincing naturalistic rationale for why the heavens should be able to answer a question from their arrangement at the time of asking pointed to horary astrology as a magical, divinatory art – and as such, in the opinion of most divines, one condemned by Scripture.

(In addition, it was widely held to be the invention of Arab infidels – somewhat misleadingly, in view of its much older and very similiar Greek antecedents.)[24]

By contrast, nativities – viewed as an examples of astral influence – seemed to retain some of the law-like determinism of natural astrology. That was not only the thrust of Ptolemy's presentation; it was also the opinion, however disputed by their clerical critics, of many early modern astrologers. To quote one,

> pretty it is to observe, how a Child, as soon as it draws breath, becomes Time-smitten by the Face of Heaven; and receives an impression from all parts of Heaven, and the Stars therein, which taking rise from the Ascendent, Sun, and Moon, and other significant places, does operate as the Impressors stand in distance, nearer or farther off: and this seems to be a concatenation of many knots which untie by course, and by distant turns; and as every knot unties, different times seem to fly out, and to do their errands; and of these sometimes you shall have two or three or more lucky knots opening together, and otherwhiles as many unlucky ones.[25]

Of course, such naturalism (even put as poetically as this) could equally well land astrologers in trouble with the Church for transgressing the freedom of the human will. For obvious reasons, divines also did not take kindly to attempts (however 'natural') to deduce the fate of churches and religions from astrological considerations.[26] At the extreme natural end of the scale, therefore, astrologers courted charges of atheism as dangerous as those of sorcery and diabolism stemming from the supernatural end. The only time astrology was quite free of official suspicion, in fact, was when it was applied in a general way to general or physical phenomena, such as epidemics or the weather.

In practice, the line between natural and judicial astrology was constantly blurred. Even in theory, the difference was never clear; given the endless variations in individuals and circumstances, it was impossible to specify a fixed point separating mere astral influence on the will, via the humours, from outright determination. Yet the need for such a distinction was a vital part of the synthesis of Christian theology and Aristotelian natural philosophy wrought by Thomas Aquinas. Aquinas's solution to the conflicting demands of both systems purchased a crucial niche in mainstream medieval Christian orthodoxy. In effect, it legitimized the basis of astrology by granting the efficacy of Aristotle's descending spheres of celestial influence, but simultaneously delimited it by stressing the freedom of the will. In medieval Europe, this compromise became widely accepted; it was a regular part of the scholastic curriculum. Only a very few (such as Nicholas Oresme and Nicholas of Cusa) questioned its legitimacy and implications.

Beginning with the sixteenth century, however, astrology came under new intellectual and moral pressure both from Renaissance humanism, as prefigured by Pico's concern to save the freedom of the will from astral influences, and from the Reformation determination to force recognition of a direct and unmediated relationship between God and man. In the latter view, astrology clearly constituted a form of idolatry. Although defended by Melanchthon, it was attacked by Luther and (above all) Calvin. At the same time, the influence of the stars on natural phenomena was still almost universally accepted. One result was that in the sixteenth and seventeenth centuries, the line supposedly dividing natural from judicial astrology became heavily policed. The distinction was still useful to all sides; it enabled divines to attack most astrological practices, and particular predictions they did not like, as having exceeded the bounds of proper natural astrology. By the same token, it provided astrologers with a relatively safe haven, since predictions could be defended by describing them as probable and natural. But both sides, more than ever divided, became committed to an interminable struggle to push the boundary in one direction or another – a contest which neither side could win outright, although (unsurprisingly, given their relative strengths and resources) astrologers were mostly on the philosophical defensive. Both the boundary and the struggle over its location, of course, were predicated on the reality of astrological influences as such. From the second half of the seventeenth century, this premise itself became increasingly problematic – a development which obliged both astrologers and divines to adapt to a new intellectual universe.

Early modern astrology varied greatly in its sophistication and modus operandi. At its most basic and popular level, the important considerations were the phases of the Moon and relatively spectacular phenomena such as comets, eclipses, and conjunctions. These were visible to the naked eye, and needed no further astronomical calculation or elaboration. At the other end (popular sharing a keen interest in comets), astrology joined natural philosophy in considering cosmic phenomena as affecting all life on Earth. Somewhere between popular and philosophical astrology were the judicial astrologers, whose interpretations of the cosmos far exceeded in complexity those of the former, but were applied – unlike those of the latter – to the lives of individuals. In what follows, it is necessary to spend a disproportionate amount of time discussing their astrology, the arcana of which otherwise invites misunderstandings.

Fundamental to judicial astrology was the 'figure', which (in a concession to modern parlance) we shall call the 'horoscope'. Astronomically speaking, this was simply a map of the planets' positions for the moment concerned, as seen from the place concerned. A particular place and time (however it is determined) are thus indispensable minimum requirements. Since the locus of the horoscope was a place on Earth, the displacement of a geocentric conception of the universe by a heliocentric required no changes in either the erection of the horoscope or its interpretation.

The planets were precisely situated in the figure in two ways: in signs of the ecliptic, or zodiac (measured from the vernal equinox at 0° Aries); and in houses, which are also divisions of the ecliptic but based on the Earth's diurnal (daily) rotation. Tackling the technical details would not be helpful here, but the following points can be noted. The zodiacal signs should not be confused with the constellations that form their starry backdrop, despite the shared names – Aries, Taurus, Gemini, Cancer, Leo, Virgo, Libra, Scorpio, Sagittarius, Capricorn, Aquarius and Pisces. The signs in Western astrology have nothing to do with the stars; they are derived purely from the Sun–Earth relationship. Seen against the constellations, they are moving very slowly backwards (taking about 25,800 years to complete a full cycle). This phenomenon, stemming from the wobble in the Earth's axis and known as the precession of the equinoxes, has been a perennial criticism of astrology (e.g. 'the signs are no longer where they used to be'). Like heliocentrism, however, precession is tangential to horoscopic interpretation, since the signs are independent of the stars or constellations. Astrologers have therefore, equally perennially, remained unimpressed by this criticism. (This is not to say that it did not figure importantly in other contexts, such as astronomy, and the perception of cycles in history.)

Regarding the houses, there were two points or areas whose importance astrologers universally agreed upon: the 'Ascendent', or horizontal angle, and the 'M.C.' (for Medium Coeli), or vertical angle. Apart from this consensus, however, the 'correct' method of house division has been disputed among astrologers for centuries. In the second half of the late seventeenth century, English astrologers could choose between at least four methods, named after their originators: the so-called Ptolemaic, which simply assigned the same degree of the sign on the ascendent to the cusp (or beginning) of each subsequent house, and the more sophisticated systems of Porphyry, Campanus, and Regiomontanus. As we shall see, the introduction of a new system by Placidus, late in the century, aroused intense controversy. In all cases, the mathematics involved were complex, but in the sixteenth century tables became widely available which permitted the necessary trigonometry to be taken for granted.[27]

The final astronomical calculation needed was that of the aspects, or the degree of angular separation (measured along the ecliptic) between each planet. Traditional (Ptolemaic) astrological theory held five aspects to be significant: the conjunction (0° separation), the sextile (60°), the quartile or square (90°), the trine (120°), and the opposition (180°). These aspects were permitted a certain inexactitude (usually a few degrees).

This was the astronomical or formal part of judicial astrology. We now come to the matter of interpretation – less straightforward, but more important. As Ann Geneva has correctly pointed out, 'Precise knowledge of geocentric astronomy was crucial in calculating the initial figure, but the true skill of the astrological practitioner resided in interpretation.' That skill consisted of translating the symbolic system of astrology into the vernacular, in a meaningful way. Given the extent to which judicial astrology became the knowledge (or if you prefer, science) that dares not speak its name – and remains so, for modern heirs of the divines, scientists and intellectual castes who found against it in the late seventeenth century, for reasons that are in part the subject of this book – it is very difficult for us to recognize this as a real skill. For its historical appreciation, however, that is what is required.[28]

I intend to mention only the most basic rules, common to all judicial astrology. They quickly develop into an intricate and arcane body of doctrine, much of which varied in its application between astrologers. To begin with, each planet was considered to have a unique nature and meaning. The basic qualities were conceived as a particular combination of masculine or feminine, hot or cold, and wet or dry. From this was derived a considerable amount of further detail pertaining to the planets' associated Earthly phenomena – temperaments, physiques, professions, illnesses, colours, herbs, weather, etc. In this way, each planet was held to signify or 'rule' an extensive range of phenomena, the connection being an agreed appropriateness of symbolism. (Thus, for example, the Sun 'ruled' and thus related gold among metals to the lion among animals, the heart among organs, the King in the civil sphere, and so on.)

The meanings of the planets were then qualified and specified in three basic ways. The most important in early modern practice was to consider what house each planet occupied. The houses were organized in four groups of three (angular, succeedent and cadent). In addition, each had its own character, relatively friendly or hostile to each of the planets, thus providing their 'accidental dignities and debilities'. Perhaps most importantly, the houses provided crucial circumstantial guidance. (Thus, for example, Saturn in the seventh house indicated delays or hindrance to marriage and partnerships; but in the ninth, to long-distance voyages, and so on.) The second way involved the signs. The twelve signs were organized in overlapping groups of four elements (fire, earth, air and water) and three qualities (cardinal, fixed and mutable), and each was ruled by a planet. Thus, each sign had a distinct individual nature, and the nature of each was again considered to be felicitous or otherwise for particular planets, providing their 'essential dignities and debilities'. (Thus, for example, the Sun was considered to be well-placed in Aries, but badly in Libra, and so on.) The third refinement was that of aspects between the planets. Basically, those based on division of the ecliptic by an odd number, such as the trine and sextile, were considered fortunate, but those based on an even number, such as the quartile and opposition, unfortunate. (Thus, Mars in quartile to another planet would indicate disputes and violence; Venus thus badly aspected, fickleness, and so on.)

These procedures by no means exhausted the repertoire. One could further consider such planetary conditions as lord of the geniture, peregrine, cazimi, void of course, and a host of others – fine points, but potentially (for the astrologer) decisive factors. It was also possible mathematically to advance a horoscope in time to any future date, in order to predict events or conditions then, or to see when a potential event indicated by the original horoscope would come to pass. The techniques for doing so were called 'directions'.

Finally, some general points about astrological interpretation should be stressed.[29] The tenets of astrological reasoning, which are developed by selective analogy, have been described as irrational. Logic of whatever kind, however, contains no guarantees as to the correctness of premises; and premises (especially assumptions) that appear arbitrary, or even irrational, to people of one age, place or class may appear just the reverse to those of another. 'Rationality' in this context is therefore an historically variable construct, signifying acceptability or plausibility. Astrological premises may lack the latter (for everyone?), but that cannot be considered an inherent defect, pointing to the irrationality of the whole enterprise.[30]

There is an apparent paradox respecting the degree of certainty (that is, degree of consensus) possible in astrological delineation. On the one hand, it was not an arbitrary procedure; a horoscope in which Saturn is the dominant planet, for example, will be found so by almost any astrologer whose competence is accepted by the majority of his or her fellows.[31] In other words, it is false to maintain that the astrologers could see in the horoscopes, and therefore predict or predicate, 'anything they pleased'; in the experience of the astrologer, there were real internal constraints. On the other hand, these constraints were never completely determining, so the horoscope was never interpretively closed. Within limits, its complexity, the lack of close consensus about which rules take precedence over others, and human ingenuity always allowed a horoscopic interpretation to be challenged or refined. As a corollary, it was also always possible for astrologers to blame the more blatant mis-predictions on (in Ptolemy's words) 'the mistakes of those who are not accurately instructed in its practice, and they are many, as one would expect in an important and many-sided art.'[32] Nor does the last point, in itself, clearly distinguish astrology from science, and establish the former as a pseudo-science, or even a 'mere' art. The under-determination of theories by facts (or by data) is now well recognized in the philosophy of science, as is the importance in scientific prediction of the condition 'all other things being equal'. And in their actual practice, scientists are far quicker to blame the application of a theory and/or the competence of its appliers, than a favoured theory itself.[33]

Finally, non-astrologers might easily underestimate the potential richness and elegance of astrological symbolism in the hands of a skilled practitioner. This was one of astrology's principal intellectual and aesthetic attractions. Yet in the case of judicial astrology, its application, given the nature of the demand from the public, was often to questions of a highly pragmatic nature. In a parallel way, astrological practice combined objective elements – the positions of the planets at a moment in time – with the different subjectivities of the astrologer's (constrained) interpretation, and the most personal and sensitive concerns of the client. For these reasons – being neither art nor science, but partaking of both – the practice of judicial astrology is best thought of as a craft.

PART I

From Heyday to Crisis: 1642—1710

2

Astrology in the Interregnum

Halcyon Days

In what way were the years of the Civil War, Commonwealth, and Protectorate halcyon days for English astrology, and why? The answer to the first part of the question is quite simply that it reached unprecedented heights of popularity and influence. This can be inferred from political and cultural events, and from the production and consumption of astrological literature. Equally remarkable were the relative independence and prestige enjoyed by judicial astrologers; here too, there are testimonials from both them and their critics.[1]

The reasons for this success lay in the potential of contemporary astrology to answer many different needs, and its acceptability as a means of doing so, together with the unprecedented availability of cheap and easy printing, in the extraordinary circumstances of breakdown in political and ecclesiastical control. Official censorship collapsed in 1641. The result was an explosion of new publications, on every conceivable subject and in every possible vein. Before 1640, there were no printed newspapers; by 1645, there were several hundreds. The publication of pamphlets now averaged three a day for the next two decades (but at a much higher rate between 1642 and 1649). George Thomason, trying to collect every book or pamphlet published, purchased 1,966 titles in 1642; in 1640, he had managed to find only twenty-two.[2]

Astrology was an inseparable part of this heady mixture, and its surviving literature not only provides the basis for any reconstruction of past astrology, it also serves as a good introduction. Until 1641, firm control of the authors and contents of almanacs had been exercized by the Company of Stationers, in conjunction with the ecclesiastical authorities,

Cambridge University and (after 1632) Oxford University. The Company's monopoly had been granted by James I in 1603, with the stipulation that

> All conjurers and framers of almanacs and prophecies exceeding the limits of allowable astrology shall be punished severely in their persons. And we forbid all printers and booksellers, under the same penalties, to print or expose for sale, any almanacs or prophecies which shall not first have been seen and revized by the archbishop, the bishop (or those who shall be expressly appointed for that purpose), and approved of by their certificates, and, in addition, shall have permission from us or from our ordinary judges.[3]

It should not be thought that the authorities were uniquely suspicious of astrology; in the 1630s, for example, their mistrust extended to every form of native or imported printed matter, not excluding ballads, and corporal penalties could be exacted on anyone who broke the decrees.[4] But it clearly included astrology. (And how right they were to be proved, from their point of view, by the 1640s and 1650s.)

This was the system which was thrown into disarray by the events of 1640–1, especially the abolition of the High Commission and the Star Chamber. The Company continued to produce almanacs, but its monopoly and control of contents broke down completely, and attempts to reimpose it in 1649 were largely ineffectual. Furthermore (and of particular importance for almanacs), in 1643 Parliament appointed the astrologer and Roundhead partisan John Booker (1602–67) as official licenser of books on 'Mathematicks, Almanacks, and prognostications'.[5]

A complete catalogue of seventeenth century works of astrology has yet to be compiled, but the estimate of a great increase after 1642, peaking around 1650, seems confirmed by works in the British and Bodleian Libraries.[6] One sign of the changed times was the appearance in 1650, posthumously, of Sir Christopher Heydon's *An Astrological Discourse*. Heydon had been a renowned astrological theorist earlier in the seventeenth century. His book was now seen through the press by a phalanx of contemporary astrologers, including William Lilly. Nicholas Fiske, in his Preface, attributed the delay in publication to 'the error or rather malice of the Clergy, who only had the privilege of licensing books of this nature'.[7]

Soon afterwards, a number of new and comprehensive astrological textbooks appeared, written in English. Previously this kind of literature

20

had been dominated by texts in Latin, occasionally translated, by Continental authorities. That the new textbooks drew heavily on older literature does not alter the novelty of their publication, as a sign of a new native robustness. The first such book was William Lilly's *Christian Astrology* (1647); it was followed by others by Gadbury, Eland, and Coley. Another major area of activity was astrological medicine, or physic; here too, the first English handbook appeared: Nicholas Culpeper's *The English Physitian* (1652). This was followed by others, by the same author as well as Saunders, Blagrave, and Salmon (whom we shall also discuss below).[8]

But the commonest publication by astrologers, which consumed most of their energy and met most of the demand (apart from personal consultation), was the almanac. The English almanac has been exhaustively studied by Bernard Capp.[9] Suffice it to say here that it was, after the Bible, by far the most popular kind of literature in seventeenth-century England. An almanac typically combined three elements: a calendar (including church festivals, markets, and fairs, along with a brief and highly selective chronology of world history); information on the year's astronomical phenomena (varying in complexity, but including at least the signs and phases of the Moon, the ingress of the Sun into each sign, and major events, such as eclipses); and astrological prognostications. Besides prognoses for the weather, crops and health, this last category nearly always embraced political and religious predictions (as discreetly worded by the author as thought necessary to avoid prosecution).

There certainly seems to have been a ready market for the new almanacs. Definite overall figures only begin in 1664, but we know that by the 1660s, average sales were about 400,000 copies a year, or roughly one for every family in three. For the seventeenth century as a whole, Thomas has remarked that the figure of a minumum of three to four million copies is 'a distinct under-estimate'. Of individual titles, Lilly's *Merlinus Anglicus*, one of the most popular, sold 13,500 copies in 1646, 17,000 in 1647, and 18,500 in 1648. Also in 1648, his *Astrologicall Predictions of the Occurrences in England* sold out four printings, totalling 6,500 copies. By 1649, his almanac was reported to be selling nearly 30,000 copies a year. And in addition to these, a large number of 'traditional prophecies' also appeared for the first time in print.[10]

Undoubtedly the unique conditions of uncertainty in the 1640s and 1650s stimulated the demand for astrology in those decades – uncertainty that embraced the metropolis and events of national importance, over and above the constant quotidian problems of rural life. Simultaneously, the unprecedented flowering of unorthodox religious and political sects provided another crucial part of the context for the astrological apogee.[11] Among radicals and mainstream alike, among the movers of events (if not quite a secure elite) and those more moved (and these very categories were in the process of negotiation), astrology found a large and sympathetic constituency. This is not surprising: it offered a traditionally sanctioned way to discern a pattern, to extract relevant meanings from the chaos of current events, and to discern future ones. In the first place, the heavens promised to reveal the divine opinion of a particular religious or political position, in circumstances where it was almost impossible not to adhere to one or another such view. Secondly, they offered a means of divine (but itself disinterested) advice in making critically important decisions, whose intangibles and imponderables ruled out arriving at any simple or obvious conclusion.

Naturally, in both cases the interpretation of the stars was rarely, if ever, disinterested; but neither was it necessarily cynical or propagandistic. The reason – and indeed what made this whole situation possible – lies in the contemporary historical conditions; at a time when natural, civil and religious phenomena were still widely seen as intimately connected, if not indeed one, there was nothing 'inherently' implausible about such a recourse. In the experience of most people, politics, religion and eschatology were ultimately inseparable; and astrological prophecy was a common thread through all.[12] As we shall see, the life of William Lilly, the leading astrologer of the time, perfectly illustrates that unity.

This is not to argue for either the primacy or purity of contemporary judicial astrology. In the welter of contending interests and values, almost no major astrologers or almanac-writers occupied neutral positions. They ranged from the fringes to the centre, on both sides of the basic divide between Anglican Royalists and Puritan Parliamentarians; and both within and without the various sects. (As we shall see, however, there were certain emphases in their distribution, and even stronger ones in the general perception thereof.) Like nearly everyone else, astrologers also occasionally fell foul of the many pitfalls; ingenuity was a vital asset in staying at liberty or alive, as were friends in high places. The direct influence of astrologers on events is very difficult to assess, but it may have been considerable on occasion, whether direct (through advice or suggestion), or indirect (through actions taken to counter or block their suggestions).

22

Astrologers, Radicals, Enthusiasts

Let us look at some of these astrologers, in the context of mid-century English society. John Booker has already been mentioned as Parliament's licenser of astrological books. Booker engaged in bitter disputes in print with the Royalist astrologer George Wharton. His almanacs contained bloodthirsty prophecies concerning the fate of Royalists, priests, and Catholics (especially Irish Catholics), and he clung to apocalyptic millenarianism when even Lilly drew back.[13]

Equally famous (notorious, to his foes) was Nicholas Culpeper (1616–54). Culpeper was wounded while fighting in the Parliamentary army, which may have hastened his early death. He was an implacable opponent of monarchy and Royalists – as evidenced, for example, in his *Catastrophe Magnatum: or, The Fall of Monarchie, A Caveat to Magistrates, Deduced from the Eclipse of the Sunne, March 29, 1652* – and of established religions and priests, proclaiming that 'all the religion I know is in Jesus Christ and him crucified, and the indwelling of the Spirit in me'.[14]

But Culpeper was best known in the context of another tradition. From the village cunning-man or -woman to the professional, 'physic' or medicine was an integral part of standard astrological practice in the seventeenth century. Conversely, astrology – sometimes in diagnosis, other times in prescription – was still a strong presence in the standard medical practice of the time. In both respects, at the time Culpeper lived, astrological physic still just spanned both the popular and elite ends of the socio-intellectual tradition. It was typified in this by his predecessors, Richard Forster (1546–1616), Simon Forman (1552–1611), and Richard Napier (1559–1634).[15]

Throughout his life, Culpeper produced a pioneering series of books on astrological, herbal and Paracelsian medicine, as well as the first English translation of the London College of Physicians' Latin dispensary. Their aim is well conveyed by the full title of his principal book: *The English Physitian, or an Astrologo-Physical Discussion of the Vulgar Herbs of this Nation . . . whereby a man may preserve his body in health, or cure himself, being sick, for three pence charge, with such things only as grow in England.* Published in 1652, it went through a remarkable number of editions, including twenty-four in the following century.[16] The English dispensary appeared in 1649, entitled *A Physical Directory, or a Translation of the Dispensatory Made by the College of Physicians of*

London, And by them imposed upon all the Apothecaries of England to make up their medicines. For thus opening their secrets to the people – few of whom could read Latin – Culpeper earned the undying enmity of the College of Physicians, whose voice was joined with those of his other critics. The Royalist news-sheet *Mercurius Pragmaticus*, for example, attacked him as a sectarian and 'absolute Atheist', accusing him of mixing every prescription 'with some scruples, at least, of rebellion or atheisme.'[17] In contemporary medicine, 'purely' medical considerations were intimately bound up with social and political ones, as in virtually every other field of knowledge or discipline. (No more so though than today, beneath the ideology of 'neutral' scientific knowledge that was established partly in reaction to exactly the events we are discussing here.)

For Culpeper, astrology was definable as simply 'that part of Natural Philosophy which inquires after the Causes, properties, Nature, and Effects of the Starres.'[18] While such an understanding was of a piece with his 'atheism' and republicanism, it should not lead us into perceiving secularism avant la lettre. For Culpeper, as for Paracelsus, nature was infused with divinity, spirituality and magicality, between which he had no interest in closely distinguishing. Natural philosophy, as the study of nature, was therefore a study of God's living book, and inseparable from true religion. And the best guides, for Culpeper, were the messages to be found in that book's macrocosmic or celestial (astrological), and microcosmic or earthly (for example, herbal), pages. Equally, in this view, 'experiment' and 'experience' (often used interchangeably) were only reliable in so far as they embodied divinely inspired insight. This kind of naturalism-cum-supernaturalism did not yet seem necessarily a self-contradiction, even for the social and institutional elites who found it so threatening. That they did so, however, is understandable. If it triumphed, not only would social chaos apparently ensue, but they would be on a level with everyone else; and what need would there be then for elite interpreters (religious, political, or medical) of God's will? Everyone would be able to interpret unaided; or, what amounts to the same thing, would think themselves able to do so.

Culpeper worked in a very long tradition of astrological physicians. His immediate successors included Richard Saunders (1613–75) and Joseph Blagrave (1610–82). Both men were friends of Elias Ashmole. Although less overtly political than Culpeper, they published books in the same tradition: the former's *Astrological Judgement and Practice of Physick* (1677), and the latter's *Astrological Practice of Physick* (1671), *Introduction to*

Astrology (1682), and an enlarged edition of Culpeper's original *English Physitian*.[19]

In eighteenth-century England, Culpeper's populist astro-physic was further extended by men such as Moore and Salmon, before finally retreating to the provinces and villages. In revolutionary England, however, socially radical mysticism was still in full flood, and it manifested itself in other astrological forms besides that of popular medicine. An esoteric astrology, embodying an illuminationist epistemology like that of Culpeper, appeared in the 1649 translation of Valentin Weigel's *Astrologie Theologized*. In 1656, John Sparrow, a Civil War officer and mystic, produced a translation of Jacob Boehme's Aurora, *'That is, the Day-Spring . . . That is The Root or Mother of Philosophie, Astrologie and Theologie'*. Jacob Boehme, or Behmen (1575–1624), was a controversial German mystic, whose writings were widely disseminated. His ideas on astrology were adopted not only by the avowed Behmenists but also by the Familists – 'a powerful influence upon the Quakers', according to Thomas, 'and barely distinguishable from some of the Ranters'.[20] The Familists' heresies were several. They included perfectionism, or the belief that it is possible to attain heavenly perfection on Earth, and mortalism, or denial of the resurrection of the human body. For this, combined with their political anarchism and proto-pantheism, they were reviled by mainstream divines. In the scandalized words of one, William Rowland, writing in 1651, 'what is THEIR Christ, but the appearance of God in every creature; a Dog, or a Cat, by consequence.'[21] Significantly, Rowland specifically identified Behmenism as the religion of the astrologers. The identification of the latter with radical sectarianism – which became widespread, virtually a consensus, a few years later – was to have momentous consequences for their future.

Lilly defended Boehme, a fact which points to the sympathy of his own astrology with the mystical intuition of the latter.[22] It is unlikely, however, that Behmenist astrologers had much time for him. This was the extreme neo-Platonic (and more specifically Hermetic) form of English astrology, as distinct from its naturalistic, Aristotelian and (later) 'scientific' forms. In chapter 25 of the Aurora, 'Of the whole Body of the Stars Birth or Geniture, that is, the whole Astrologie', Boehme wrote of astrology and its present masters that

indeed it has a true Foundation, which I know in the Spirit to be so, but their Knowledge stands only in the House of Death, in the outward

Comprehensibility or Palpability, and in the beholding with the Eyes of the Body . . . but my knowledge standeth in this Birth or Geniture of the Stars, in the Midst or Center, where Life is generated, and breaketh through, and in the impulse or moving thereof, I also write.[23]

However obscure or woolly-minded such a passage may sound today, such a perception should not be allowed to confuse the issue of how it appeared in the 1650s. Behmenist astrology, which the Familists were confident would help them to 'conquer over the whole world', was one of several such heterodoxies whose worrying connotations for the authorities (whether Royalist or Cromwellian) were equally political and religious.[24] From this point of view, the most cogent contemporary fact about astrology was its ubiquity in the upheavals of the times – a presence clear not only from the high profile of eminent astrologers like Lilly, Booker, Culpeper, and the ominous if more nebulous advocacy of Familists, but also from the well-documented recourse to astrologers by radicals of all kind: Fifth Monarchists, Levellers, Ranters, and Diggers, down to mere republicans and Civil War officers.[25] Among Fifth Monarchists, John Spittlehouse had high praise for astrology, and for Lilly as 'the prince of astrologers'; Owen Cox consulted Gadbury and Lilly; and Peter Chamberlen, later a Seventh Day Baptist, was sympathetic to astrology.[26] The leading Leveller Richard Overton (and, as Thomas notes, as 'rationalist' a man as any of the time) consulted Lilly in April 1648, as to 'whether, by joining with the agents of the private soldiery of the Army for the redemption of common right and freedom to the land and removal of oppressions from the people, my endeavors shall be prosperous or no'.[27] Mrs Lilburne, the wife of the Leveller Freeborn John, regularly consulted Booker, as did William Rainsborough, a Civil War officer, Leveller, and, later, Ranter. Laurence Clarkson (or Claxton), who also made the transition from Leveller to Ranter, reflected in his autobiography on his own short career as an astrologer:

> now something was done, but nothing to what I pretended; however monies I gained, and was up and down looked upon as a dangerous man, that the ignorant and religious people was afraid to come near . . . yet this I may say, and speak the truth, that I have cured many desparate Diseases.[28]

Yet it should not be thought that astrology among the radicals was just, in the words of a contemporary, 'a study much in esteem among illiterate Ranters'.[29] None of the men just mentioned were illiterate; in particular, besides Overton, there was the remarkable Gerrard Winstanley, the

26

leading Digger radical and pamphleteer, who advocated that astrology should be taught in his utopian society.[30] A similar view was recorded by the Puritan radical John Webster (1610–82), in his *Academiarum Examen* (1654). Webster attacked the backward-looking conservatism of the universities, especially their Aristotelian syllabuses. His prescription for this state of affairs mixed Baconianism and the new science with alchemy, Paracelsian medicine and astrology. The latter was deemed as an art 'high, noble, excellent, and useful to all mankind', and an accompanying paean praized his 'learned, and industrious Country-men Mr. Ashmole, Mr. William Lilly, Mr. Booker, Mr. Sanders, Mr. Culpepper, and others, who have taken unwearied pains for the resuscitation, and promotion of this noble Science'.[31] John Wilkins and Seth Ward – later founding members of the Royal Society, and Anglican Bishops – responded to Webster with anxious alacrity. Nor, finally, were astrologers only praised or consulted by the relatively extreme or powerless radicals. Prominent Parliamentarians, as we shall see, protected Lilly.[32]

My point here is neither to demonstrate the political progressiveness of astrology at this time, nor even to show that the two were perfectly compatible (a point true as far as it goes). It is simply to emphasize the deep involvement of astrology and astrologers with mid-century radical politics and religion, with a view to the crucial historical consequences of that involvement for the former.

There were also astrologers on the other side of the chief contemporary divide. The outstanding and highly courageous opponent of Booker and Lilly, during the Civil War, was George Wharton (1617–81).[33] Imprisoned in 1649–50, and in danger of losing his life, Wharton was released through the intercession of his friend Elias Ashmole, who in turn prevailed upon Lilly to use his influence among the Parliamentarians. Lilly did so, which undoubtedly helped his own cause after the Restoration (apart from the personal satisfaction of showing magnanimity to an old rival). Wharton was appointed Royal Paymaster after 1660, and made a baronet in 1667. He was a close friend of Ashmole and Sir Jonas Moore, both Fellows of the Royal Society. Although continuing to pursue his astronomical interests, he ceased producing an annual almanac after 1666.

This biography, though brief, is sufficient to show the very different social character of Wharton, the Royalist astrologer par excellence, from that of his leading Parliamentarian rivals. Even Anthony à Wood, no fan of astrologers, but loyal to his own, declared that 'Sir George was always esteemed the best astrologer that wrote ephemerides of his time, and went beyond William Lilly, and John Booker, the idols of the vulgar'.[34] It also seems significant that there were no other astrologers of even roughly equal stature on the same political side; an exception might be John Gadbury, but he appeared on the scene later and in very different circumstances, after 1660.

William Lilly and 'Democratic' Astrology

Two men in particular acted as the focus for judicial astrology in mid-seventeenth century England: William Lilly (1602–81) and Elias Ashmole (1617–92). From them, the nexus spread out to take in virtually the entire active astrological community.[35] The one who commanded most attention from his contemporaries was Lilly.[36] Born in the Leicestershire village of Diseworth and educated in the local grammar school, he came to London to earn his livelihood at the age of seventeen. In the course of a variety of pursuits, the chance reading of an almanac aroused his interest in astrology. He began to study it in earnest from the age of thirty, including a spell of tutelage under John Evans, a 'cunning man' residing in Gunpowder Alley. He started to attract clients from about 1635, and the success of his first and subsequent almanacs has been mentioned.

Lilly had the right touch for the time and place. He mixed traditional forms, for example *A Prophecy of the White Kings Dreadfull Dead-man Explaned* (1644), with more astrologically precise judgements on the major issues of the day, such as resolving 'If Presbytery shall stand?' from the detailed interpretation of a horary figure. (His hopeful conclusion was that 'the Commonalty will defraud the expectation of the Clergy, and so strongly oppose them, that the end hereof shall wholly delude the expectation of the Clergy'.[37] His political sympathies, openly displayed, were moderate Parliamentarian; and his (correct) predictions of defeat for the Royalist forces at Naseby (in 1645) and three years later at Colchester, made him as popular among the former as he was hated by the latter. The same year as he was cheering on the besiegers of Colchester, however, he was consulted (probably for a second time) by an emissary of Charles I, Lady Jane Whorwood, on how best to escape from Carisbrook Castle in the Isle of Wight. Ever practical, Lilly supplied both advice and a saw, although to no avail. He seems to have been genuinely horrified by Charles's beheading in 1649.

In that, he may not have been very different from many of his supporters. These came chiefly from the lay Independents and army radicals, and featured men such as Bulstrode Whitelocke (1605–75), a Keeper of the Privy Seal and Commissioner of the Treasury, whose life he regarded Lilly as having saved through the latter's astro-medical intervention during a serious illness.[38] In return, Whitelocke more than once saved Lilly from his enemies. These remained numerous, even after Cromwell's triumph, and consisted in the main of Presbyterians. At their instigation, he was examined by Parliamentary committees in 1645 and 1652, and twice briefly imprisoned. Lilly's politics were thus on the radical side, but neither uncritical – he faulted Parliament on more than one occasion – nor extreme. Despite his admiration for the writings of Boehme, he was unsympathetic towards Fifth Monarchists, Levellers, and 'that monstrous people called Ranters'.[39]

In 1652, Lilly withdrew from London to Hersham, in Surrey. He continued producing almanacs, and kept his old 'Corner House on the Strand' for his busy consultation practice. Thomas (whose description of astrology in the consulting room remains unsurpassed) has estimated that Lilly's clients were nearly equally divided between men and women; about one-third may have been female servants, but a high proportion of the remainder were 'gentry or above'. Questions covered all possible aspects of personal life, from children and love-life, to medical problems, to questions of political allegiance. His questioners wanted information or advice, and usually the more precise the better. For this service, Lilly charged a standard rate of about half a crown (12½p) but nothing for the poor, and as much as £40–50 for those members of the aristocracy who could afford it. He also taught astrology for a fee, and for some years received a state pension. All this brought him an income well towards the top end of the contemporary scale (in which he was untypical of most of his colleagues).[40]

Lilly stands out in other ways, too. He was consulted by MPs, members of the aristocracy and some leading radicals. He was generally admitted among judicial astrologers to be their leading representative, and Abraham Whelocke, a professor and Head Librarian at Cambridge University, was accurate in hailing him as the chief 'promoter of these admired studies'. Even his enemies paid him the compliment of complaining that the people 'put more confidence in Lilly than . . . in

God'. John Evelyn recalled in 1699 that during an eclipse 'fifty years ago many were so terrified by Lilly that they durst not go out of their houses.' This degree of influence – even more than his record of alarming predictions in matters of State and religion – brought him to the attention of the authorities, both before and after the Restoration.[41] Ironically, in so far as Lilly was identified as the leading representative of his profession, it also helped to undermine the position (and thus eventually, the influence) of his successors.

In his astrology too, Lilly is worth noting as someone whom subsequent developments would render unusual. I am not just referring to his authorship of the first English textbook of astrology, nor to its quality and thoroughness.[42] Lilly's practice relied heavily on horary work (that is, the openly divinatory kind of astrology which is used to answer questions based on a figure of the heavens at the moment the question was asked, or received). There was a strong practical argument in favour of horaries, as distinct from nativities: many people did not know their precise time of birth. Nonetheless, horary practice was criticized by many, including some astrologers, as indefensible on grounds of reason or religion; and it was a favourite target of divines. But Lilly's attitude towards astrology en tout was strangely straightforward and unproblematic: 'the more holy thou art, and the neer to God, the purer Judgement thou shalt give'.[43] Lilly's God thus permitted astrological predictions, of a kind akin to prophecies, vouchsafed to the pure. At the same time, however, He forbade hubris; every extant portrait of Lilly shows him holding a horoscope bearing the words, *'Non cogunt'* – that is, in the words of the old maxim, 'the stars incline but do not compel'. There is no reason to think that this functioned for Lilly and his peers purely as a cynical escape-clause, though doubtless it was that for some; in Lilly's case, his personal (if idiosyncratic) piousness is evident from his writings. But if this attitude to astrology now seems eccentric, it is for two good historical reasons: because it embodies a 'pre-Enlightenment understanding', in which the knowing subject and the objects of knowledge are not seen as cleanly separable;[44] and because it successfully brings off an amalgam of Christianity and popular magic. Generations succeeding and even surrounding Lilly, overtaken by the effects of the Reformation and Counter-Reformation, were persuaded (if that is the word) that these two were fundamentally incompatible.

Lilly seems to have been unaware of, or at least unconcerned by, any such incompatibility. In the early 1640s, he dabbled briefly in overtly magical practices, such as summoning spirits, before giving them up and burning his magic textbooks. But his astrology never lost its divinatory, implicitly magical quality. Indeed, he attributed his more spectacular predictive successes (of which more later) to 'the Secret Key of Astrology, or Prophetical Astrology'. Unlike the leading English astrologers of the next generation, John Gadbury and John Partridge, he neither showed the slightest interest in a 'scientifically' purified astrology, nor agonized over whether his astrology was rational or not. Lilly was really the last great English astrologer who could unselfconsciously advocate astrology as a divinatory craft, and issue his predictions as prophecies.[45]

The reasons for this fracture in the astrological tradition are, in large part, the subject of this book. What is important to note here about Lilly is that his approach was (like that of Culpeper) both magical and demotic, or democratic. As such, it was characteristic of much Interregnum radical thought. And these two strands were indissolubly linked: guidance concerning the microcosm (the Earth) was to be found in the messages of the celestial macrocosm, the coherence of which was guaranteed by relations of sympathy and antipathy; and these messages could be read by anyone with sufficient application to acquire the necessary skills, and the requisite attitude. Hence Lilly's textbook, which attempted to disseminate these essentials. But whether one possessed skill and holiness was ascertainable principally by the success of one's astrological predictions, not by any human authority. Furthermore, Lilly's maxim, 'Non cogunt', not only left him with room for manoeuvre; it also granted his audience that freedom. It would have been quite inconsistent for him to have adopted a strongly determinist position.

In social terms, Lilly can therefore be seen as a spokesperson not for popular culture – for there was always a high degree of ambivalence in the relations between the latter and radical thought – but for its radicalized and self-conscious elements, whose democratic demands in the mid seventeenth century proved almost as uncomfortable for Cromwell as for the Stuarts (although more completely and successfully rejected, from 1660, by the latter).

Early modern science has sometimes been credited with destroying astrology. Certainly it did contribute to astrology's 'death' after 1660, but not in any direct or obvious way, as Lilly's circle of friends and colleagues demonstrates. They included some of the most advanced astronomers in the country. To some extent, the distinction between astrologers and astronomers is a modern, anachronistic one in relation to the seventeenth century. Such a distinction was available then, but it was still in the process of negotiation, more a matter of contention than consensus. On the one hand, the intellectual stimulus of new discoveries and theories by Brahe, Kepler, and Galileo were as exciting for philosophically inclined astrologers as they were for astronomers. (Against the ready tendency to read that word in a purely modern sense, which excludes astrology, it is salutary to recall that the first two of those three astronomers were themselves deeply committed to particular kinds of astrology.)[46] On the other hand, the institutionalization of that work after 1660, especially under the aegis of the Royal Society and through the work of Flamsteed, Halley, and Newton, was proceeding in a way that (as we shall see) effectively excluded any overtly astrological considerations from its domain. In mid-century, however, the overtly astrological significance of the new discoveries and theories was still a question as open to natural philosophers as it was to the public.

One person who found it particularly stimulating was Jeremy Shakerley (1626–c.1655).[47] A highly competent observational astronomer, Shakerley was the second person (after Gassendi) correctly to predict and observe a solar transit of Mercury, which took place on 24 October, 1651. A member of the influential Towneley circle, he was also quick to recognize the importance of Jeremiah Horrocks's work. Shakerley has been acclaimed as a representative of the new scientific zeitgeist, a seventeenth-century modern – unlike many of his contemporaries, a 'taterdemalion army' of astrologers, alchemists, and the like.[48] If so, he is a curious choice; for when he asked rhetorically – inspired by the new astronomical advances – 'O Heavens and Stars! how much hath our age triumphed over you! . . . Why then shall we subject ourselves to the authority of the Ancients, when our own experience can inform us better?'[49] Shakerley was referring to astrological authority as much as astronomical. In a letter to a friend written in 1649, referring to what must have been a request for astrological guidance, he conjectured that

> if there be help from astrology (which I will not deny may be) I believe that not only the art but even the key of the art is locked up . . . Can we believe them [i.e. astrologers] in contingencies, [we] who are to seek in the things that must follow of necessity and depend on the constancy of Nature's powers? . . . if God spare me life and ability, I shall show in public what my thoughts are upon this subject, and perhaps from philosophical principles seek a foundation for a more refined astrology.[50]

Thus Shakerley, although severe, was far from rejecting astrology outright. And a further dimension is perceptible in the several respectful, not to say sycophantic, letters he left behind written between 1648 and 1650 to William Lilly. It was from Lilly that Shakerley had received a copy of the astronomy textbook *Urania Practica*, by Vincent Wing and William Leybourne. In response, he published a critique of its authors' grasp of lunar theory.[51] That seems to have led to a break between the astrologer and the astronomer which seriously distressed the latter man. The earlier letters reveal his attempts to cultivate Lilly's patronage, but the last one says, 'Now that you have rid yourself of mee, you need not long be separated from the friendship of Mr. Wing and Mr. Leybourne.' And he defended himself (convincingly) against Lilly's unrecorded charge of 'conspiring with [Jonas] Moore against you'. Not long afterwards, Shakerley left for the Far East; a final letter survives, posted from India, where he is presumed to have died.[52]

Vincent Wing (1619–68) is also a challenging figure in this context. It would be both rash and fruitless to say whether he was 'really' an astronomer or an astrologer. He and his scions made Rutland a leading centre for both disciplines, as well as mathematics and surveying, into the final decades of the eighteenth century. Wing himself produced several important astronomy textbooks, in which he advocated and elaborated Copernican and Keplerian concepts. He also compiled what Flamsteed described as 'our exactest ephemerides'.[53]

Wing was best known to the public, however, for his very popular annual almanac, including its astrological prognostications. This was no cynical, purely remunerative exercise (as used to be maintained, equally falsely, of Kepler).[54] In addition to his manifestly painstaking and sincere almanacs, Wing saw through the press, and wrote a preface for, George Atwell's posthumous *An Apology, or, Defence of the Divine Art of Natural Astrology* (1660). Perhaps most tantalizing, however, is his surviving correspondence. It includes, for example, a horoscope for the local appearance of 'three Suns', which he sent to Lilly for his 'Astrological iudgmt'. Another letter, dated 28 July 1650, respectfully requested Lilly's opinion on a horary figure concerning the theft of some linen from an MPs wife, and 'whether they bee recoverable or not'. (Both times, Wing himself had set the initial figure.) In the same letter, Wing also asked for 'a line of Comendacion' of his new book, Harmonicon Coeleste, in Lilly's next almanac.[55]

Despite Wing's Royalist beliefs, he also wrote in respectful terms to his

'Learned and truly honoured friend', John Booker. Another friend, perhaps his closest, was the leading Royalist astrologer and almanac-writer after 1660, John Gadbury. Gadbury wrote a eulogy of Wing after the latter's death.[56] And like all these men, Wing was a friend and regular correspondent with that other pole and chief support of the astrological community, Ashmole.

Before turning to Ashmole, a third figure demands some attention: that is, Thomas Streete (1621–89).[57] Also a firm heliocentrist, Streete made observations with both Hooke and Halley, and his method of deducing longitude from the motions of the Moon won the commendation of Brouncker, Wilkins, Pell and Moore, of the Royal Society. His *Astronomia Carolina* (1661) was a leading textbook for many years, disseminating Kepler's laws of motion to many (possibly including Newton). A second edition in 1710 was edited, with an appendix, by Halley.[58] At the same time, Streete produced several ephemerides for astrologers' use, containing both geocentric and heliocentric positions and aspects of the planets. His first, appearing in 1652, was the first such publication in England (along with that of his friend Joshua Childrey). In it, Streete recorded his belief that both kinds of aspects have great force, and 'long continued and infallible experiments do also confirm the same, both in the Change of Weather, and things of near concern to us.' Also like Wing, he too was a Royalist – the rule, when it comes to astrologers of this period who were also involved in natural philosophy – and, as befitting someone writing in 1652 who had fought for the King, he added prudently, 'Let honest Astrologers be cautious herein, as also in the rectifying of Nativities, and most in their Predictions. So will I.' (Of course, it was also possible that he was advocating caution for 'scientific' reasons – the two are not mutually exclusive – although that interpretation sits a little uncomfortably with the simultaneous attestation of 'long continued and infallible experiments'.)[59]

My object here is not to show that astrologers were in the scientific vanguard of the mid-seventeenth century. I am more concerned with the fluidity of disciplinary boundaries, and (more importantly) of ways of thought and behaviour that have since acquired much more distinct and even hostile identities: magic, or 'occultism', and natural philosophy, or 'science', astrology and astronomy, radicalism and reason. The fact that subsequent developments drew a sharp line between the former and latter terms of each pair remains a constant temptation to misinterpret and/or misrepresent their common history today.

Elias Ashmole and 'Elite' Astrology

Although Lilly was a more public figure, any consideration of the pre-lapsarian world of English judicial astrology would be seriously incomplete without Elias Ashmole (1617–92).[60] Ashmole was unique in his contacts and patronage within the astrological community. He employed Booker, Streete, and the judicial astrologer Richard Edlin as clerks in the Excise Office. He was a good friend of Wing, Gadbury, and Wharton. He was probably closest to Lilly, and paid for Lilly's handsome tombstone after his death. But he knew Booker too, and performed the same service for him. He also befriended the astrologer-physicians Blagrave and Saunders, among others.

Ashmole's politics were unswervingly Royalist, and after 1660 he held office as Windsor Herald and Comptroller of the Excise. As Michael Hunter puts it, there can be no doubting his 'lifelong commitment to the status quo and to the monarchy as its embodiment'.[61] He was also deeply enamoured of Freemasonry and hermeticism, especially alchemy, as revealed by an extensive library as well as his own writings. A founding member of the Royal Society, he was a representative figure of those (a not inconsiderable number) who saw no necessary conflict between magical knowledge and natural philosophy; the new experimental science was to purify and ennoble both. This view is evident in his definition of magic: 'the Connexion of naturall Agents and Patients, answerable each to other, wrought by a wise Man to the bringing forth of such effects as are wonderfull to those that know not their causes.'[62]

Ashmole's interest in astrology began in 1644, at the age of twenty-seven. His senior officer in the defence of Royalist Oxford was Sir John Heydon (d.1653), apparently the second son of Sir Christopher Heydon, and 'mystically inclined'.[63] In the following year, he met Wharton, and in 1646 he was introduced (by Jonas Moore) to Lilly. After some misgivings concerning the latter, he became firm friends with both men. (He did not meet Gadbury, with whom he also became close, until 1652.)

Ashmole's appetite for astrology proved insatiable. He acquired the manuscripts of the Elizabethan astrologer Simon Forman and the Jacobean astrologer-physician Richard Napier (as he was later to do of Lilly), and applied his knowledge to every conceivable circumstance and choice. The chief vehicle for this practice was a painstaking analysis of his own geniture, including directions and revolutions – the last being an

35

annual figure when the Sun returns to its original position at one's birth – along with apparently endless numbers of horary and electional horoscopes. We know that in 1648 it took Ashmole only seven to fifteen minutes to calculate a basic nativity. (Its interpretation could take as long as was practically possible.) He only started learning how to direct a figure, however, that same year – four years after beginning his astrological studies. The length of time spent studying the 'basics' is indicative of the seriousness with which Ashmole, and others like him, approached the subject. Ashmole's high opinion of it was stated simply in his *Theatrum Chemicum Britannicum* (1652): 'Iudicall Astrologie is the Key of Naturall Magick, and Naturall Magick the Doore that leads to this Blessed Stone' (that is, the Philosopher's Stone).[64]

So far, none of this strikingly differs from the astrological practices or beliefs of the other astrologers I have so far described. But there were differences none the less, which should be carefully noted. Against the view that a monopoly of magical interests was held by Interregnum radicals, it has been argued that Ashmole was one of 'an equally significant core of men who combined strong support for the Stuart monarchy with mystical world-views.'[65] And it is true that he was tapping the same broad tradition of natural magic. Closely examined, however, his astrological writings show that Ashmole emerged with a rather different set of conclusions from those of the radicals.

To begin with, compared with the illuminationist epistemology and macrocosm-microcosm cosmology of Boehme, or even Lilly, those of Ashmole strongly emphasize the importance of hierarchy, and (by implication) one's rightful place therein. Hierarchy in this view extended from inanimate nature, through a series of finely distinguished grades, to pure spirit. It was at one and the same time mystical, natural and social. Thus, to affirm its importance in one realm was to affirm it in all. It was also epistemological, since the availability (or at the least, advisability) of knowledge was importantly determined by one's relative place on the ladder.

To cap it all, the frontispiece of Ashmole's first book includes a depiction of his nativity, the centre of which bears the words '*Astra regunt homines*' (the stars rule man). Such fatalistic determinism emphasizes his one-way conception of hierarchical magical influence, and contrasts strikingly with the meliorist maxim in Lilly's portraits: '*Non cogunt*' (the stars do not compel). That quasi-egalitarian view should not be confused with Ashmole's interest in magical operations, none of which suggests that the relevent hierarchies can be subverted or avoided. (Of course, he

36

applied a rather different set of rules to himself. Highly ambitious, he had few scruples in matters of his own advancement. But that is hardly uniquely damning; the same could be said of Lilly.)[66]

Of a piece with this attitude were Ashmole's strongly anti-democratic views. He worried that whereas 'Judiciously dispens'd to the World', astrology would excite admiration, if 'unskillfully exposed' it would only become 'the scorne and contempt of the Vulgar'. He warned his readers,

> Trust not to all Astrologers . . . for that Art is as secret as Alkimie. Astrologie is a profound Science: The depth this Art lyes obscur'd in, is not to be reach't by every vulgar Plumet that attempts to found it. Never was any age so pester'd with a multitude of Pretenders . . . of this sort at present are start up divers Illiterate Professors (and Woman are of the Number) who even make Astrologie the Bawd and Pander to all manner of Iniquity, prostituting chaste Urania to be abus'd by every adulterate Interest.[67]

In its own way, Ashmole's concern was justified. But his prescription – confining initiation into astrology's secrets to a small social and spiritual elect – was quite unworkable; the situation had long since passed out of his or anyone else's control, if it had ever been otherwise. Nonetheless, the implications are plain, and they match his immersion in hermetic esotericism and Freemasonry. What could be more different from Lilly's pains with *Christian Astrology*, in which he 'omitted nothing willingly, which I esteemed convenient or fit'; or Culpeper's *English Physitian*, teaching the astrology whereby 'a man may preserve his body in health, or cure himself'?[68] The 'spirituality' of both Ashmole's and (among others) Lilly's astrologies could easily mislead a modern observer into assuming an identity thereof. At bottom, however, Ashmole's elitist and conservative values had little in common with the meliorist and democratic egalitarianism of the radicals.[69]

Ashmole's esotericism, in conjunction with sufficient income, took an appropriate social form; he never embraced any kind of popular practice, preferring to employ his astrological expertise in answering a remarkable series of questions he received after the Restoration from members of the gentry. These included Sir Robert Howard, Sir Thomas Clifford, and even Charles II (on the prospects for his delicate relationship with Parliament in late 1673).[70] Their confidence is evidence not only of the survival of magic in high places (although now in secret), but also of Ashmole's secure niche among the social elite.

To find such favour, incidentally, it was not sufficient merely to harbour Royalist sentiments. John Heydon (1629–*fl*.1667) was imprisoned during the Interregnum for using astrology to foretell Cromwell's death; he was released about 1657, only to repeat the experience – for the same reason – under Charles II, in 1663 and 1667.[71] But Heydon was a loyal subject of His Majesty, who attacked Lilly and Booker and complained that 'The recent years of the tyranny admitted stocking-weavers, shoe-makers, millers, masons, carpenters, bricklayers, gunsmiths, porters, butlers, etc. to write and teach astrology and physic.'[72] Heydon seems to have had a similiar audience in view, however. His astrology was a sort of populist poor man's esotericism, presented in books such as *The Rosie Crucian*, with 'Infallible Axiomata, or, Generall Rules to know all things past, present and to come' (1660); and *The Harmony of the World*, 'Being a Discourse wherein the Phaenomena of Nature are Consonantly Salved and Adapted to Inferior Intellects' (1662). For this perverse dalliance with populist magic – not to mention drawing attention to such an astrology at just the time when that was most undesirable – it is not surprising that Heydon fell out with Gadbury, and was described by Ashmole as 'an ignoramus and a cheate.'[73]

In terms of a contemporary division of judicial astrologers between radicals and conservatives, it could be argued (and I would) that there were fewer of the latter; but that is a quantitative impression which is difficult to substantiate. What is surely beyond reasonable doubt is their lower public visibility, qua astrologers. This was true not only of Ashmole, but of others, such as John Aubrey, Edward Dering and John Dryden. As I have tried to show, such behaviour was a concomitant of both their social positions and interests and their philosophies of astrology. It is only consistent to conclude – and in the absence of direct support, which would be difficult to imagine, then at least uncontradicted by the historical evidence – that a widespread association of astrologers with radicalism (atheism, republicanism, enthusiasm) was not a whit moderated, let alone obviated, by the relatively hidden presence of Royalist counter-examples.

There was always at least one 'public' Royalist astrologer in the field: after Wharton (who retired from active astrology shortly after the Restoration), Gadbury, and after him George Parker. But such astrologers seem to have remained in a minority compared with their Parliamentarian, and, later, radical Whig, opponents. And at first glance, Gadbury's commitment to the transformation of astrology by

Baconian methods into a more objectively (and therefore, putatively, publically) verifiable science appears to contradict Ashmole's elitism and privacy. But the contradiction is superficial. We noted Ashmole's hopes for the new science, which meshed with those of Gadbury. To anticipate, the younger man may be seen as involved in the next phase of a key conservative programme, centred in the post-Restoration Church and the Royal Society: to develop an apparently objective approach to knowledge, and thus to produce a body of knowledge that would command public consent without resorting to coercion. (Of course, this knowledge was as socially interested and selective as any other, but it was vital to conceal that in order to avoid the kind of open schisms the country had just experienced.) As we shall see, Gadbury was a leading figure in the attempt, between 1660 and about 1690, to convert astrology into a form of this new and safe knowledge, and thereby ensure its survival in a world grown sharply hostile to 'enthusiasm'. Ashmole's projected oligarchy of magicians was thus replaced by one of natural philosophers. That Gadbury's efforts failed does not invalidate the similarity of his atttude; and late in life, like Ashmole, he too regretted having exposed astrology to such public view, thus encouraging its use and abuse for various ends. It was a mistake, he felt, to have divulged 'so many of Urania's secrets to common eyes. But what is done is past retrieve.'[74]

Despite their differences in values and orientation, Lilly and Ashmole still shared not only a common language at least of natural magic, but contiguous social worlds. What is perhaps most striking about their friendship is that it was possible, pointing not so much to an admirable but eccentric ecumenism as to the set of conditions that began to disappear in 1660, after the experiences of the Revolution. In the post-Restoration world, seen through the window of contemporary astrology, a line was drawn – not between magic and reason, as is so often assumed, but through magic, dividing it into acceptable and unacceptable sorts.

The force of this social and intellectual redefinition is clearly perceptible in the differences between Lilly and Ashmole and their heirs in the following decades. It is virtually impossible to find one who openly espoused a magical or divinatory astrology. That was simply no longer a viable option, beyond the ambit of the town cunning-man or -woman. Instead, judicial astrologers divided into those claiming that the basis of their art was rational in 'scientific' terms, and those holding out for an equally rational Aristotelianism. Indeed, both sides undertook drastic and competing reforms, in an attempt to win back the lost approbation of elite culture. Quite as much as Lilly, however, the same process left Ashmole behind (together with others enamoured of natural magic in the early Royal Society) as a relic in his own time. The new post-Restoration world demanded a magic restricted to those who could be entrusted with its use, and stigmatized any other kind as vulgar enthusiasm. In this context, Ashmole's naturalistic definition of magic (quoted earlier) opened the door to a new generation of natural philosophers with little sympathy for his unreconstructed sort. The philosopher-mage whose efforts crowned the new edifice was, of course, Isaac Newton.

The Astrologers' Feasts

The Royal Society of London was preceded there by a rather different organization. The differences between the two starkly emphasize the changes after the Restoration which divided them. Ashmole's journal, supported by a few other documents, shows that during the years 1647 to 1658, a group of about forty astrologers met several times a year in London for a banquet and a sermon. This group called itself the Society of Astrologers of London. On these occasions, political and religious differences seem to have successfully been set aside. Ashmole noted the following meetings or 'Feasts': 14 February, 1647 (at the White Hart in the Old Bailey); 1 August, 1649 (at the Painters-Stainers Hall); 3 October 1649; 8 August 1650; 14 August 1651 (again at the Painters-Stainers Hall); 18 March 1653; 22 August 1654; 29 August 1655; and lastly, 2 November 1658. Assuming that he was unable to attend all the meetings, there may well have been others. At each meeting, a Steward was apparently chosen to arrange the following one; Ashmole was so delegated in 1650.[75]

Another key figure, given his stature in the astrological community, was Lilly. *A propos* a meeting unmentioned by Ashmole, Lilly recorded in his almanac for 1649: 'And I heartily salute with many thanks, all that civil society of Students, being in number above forty, at our sober meeting October 31. last: among all which number, during our continuance together, there was no one oath heard, no health in drink once mentioned, no dispute of King, Parliament, or Army.'[76] Lilly is also the only member mentioned by name in a florid letter from the Society to Bulstrode Whitelocke. This letter, dated 24 April 1650 and in the handwriting of George Wharton, thanked Whitelocke for his 'great

40

Encouragement in the past', and asked for his continued patronage. Astrology in England, it declared, was only lately restored 'by the dextirous Scrutiny and Pains of Mr. Lilly', and of course, 'your Lordship'. As a powerful figure in the Parliamentary government, and chief protector of Lilly, Whitelocke was a good choice of patron.

No lists survive of the astrologers who attended. Forty seems a reasonable number, however, and it is encouraging (assuming that at least some astrologers occasionally came from outside London), we know of about thirty who could have been there.[77] The Painters-Stainers Hall, where at least two feasts took place, still exists in its original place, in Little Trinity Lane, London EC4. The original building was destroyed in the Great Fire of 1666, however; the new building was damaged, although not destroyed, during the Second World War.[78] Some idea of a typical menu may be had from what Ashmole once served at a dinner for his benefactors in 1677:

First Course
  Haunch of Venison and Cawley Flowers
  Batalia Pie
  Ragowe of Veal
  Venison Pasty
  Chyne of Mutton and Chyne of Veal in a Dish

Second Course
  Chickens Ducks and Turkeys
  Chyne of Salmon and Soales Fried
  Kidney Beans
  Westphalia Gammon and Tongues
  Joale [i.e. jowl] of Sturgeon
  Fruit $\left\{\begin{array}{l}\text{Apricots}\\\text{Cheries}\\\text{Gooseberries}^{79}\end{array}\right\}$ severall kindes

After the Astrologers' Feast of 14 August 1651, Ashmole fell ill from a 'surfeit', occasioned 'by drinking water after venison'. As a result he felt 'greatly opprest' in his stomach. The ministrations of Dr Thomas Wharton, an eminent physician and anatomist of Trinity College, Oxford, proved to be of no avail. But the next day, Ashmole's friend 'Mr. Sanders the astrologian sent me a piece of bryony root to hold in my hand; my stomach was freed of that great oppression, which nothing which I took before from Dr. Wharton could do before.'[80] Thus did natural magic triumph over natural philosophy, on the battleground of Ashmole's stomach!

Some of the sermons delivered at the Feasts also survive. Their authors were a collection of religiously peripatetic, not to say eccentric, divines. Robert Gell (1595–1665) had preached in 1631 before Charles I, and in 1641 before the Lord Mayor of London. He seems to have become more radical during the Interregnum, and is described by Thomas as a Familist. Gell twice preached before the Society: in 1649, on the text of Matthew 2:2 and the wise men who 'knew Him by the star'; and in 1650, when he delivered *A Sermon Touching Gods government of the World by Angels* (hardly standard Puritan theology). He condemned both 'notorious Fatalists' (that is, Presbyterians) and 'the Rantists'.[81] Also in 1650, Edmund Reeve (d.1660) gave a sermon. A Laudian, Reeve had been rejected by the Commonwealth authorities. His sermon, *The New Jerusalem*, discussed the Book of Revelations – always a rich source of astrological symbolism. Arguing that there are both lawful and unlawful kinds of divination, and that astrology is properly one of the former, Reeve then listed and answered a number of standard objections to astrology.[82] In 1653, Thomas Swadlin (1600–70) – another Laudian and a Royalist – prepared a sermon entitled *Divinity No Enemy to Astrology*.[83] Lastly, Richard Carpenter (*fl.*1650–70) gave a sermon in 1657, entitled *Astrology Proved Harmless, Useful, Pious*. Sometime vicar of Poling, Carpenter seems to have changed his religious convictions frequently; at different times, he espoused Catholicism, Protestantism, and Independence. His text on this occasion was Genesis 1:14: 'And God said, let there be lights in the firmament of heaven . . . and let them be for signs, and for seasons.' Along with the example set at Jesus's birth by the three Wise Men, this was a fundamental biblical reference for astrologers in their own defence. Carpenter used it to argue that the planets and signs both signify, as signs, and transmit God's will by their influences. 'The doctrine of Aquinas,' he said, 'is impregnable, and stands like a Tower in the Fort of Reason.'[84]

These were all standard Biblical defences of astrology, many centuries old.[85] Doubtless they were as comforting to the audience of astrologers as they remained unconvincing to their many Puritan and especially Presbyterian critics. The obscurity of these divines and their rather hackneyed arguments, along with the somewhat perfunctory attitude of

42

both Lilly and Ashmole to institutionalized religion, suggest that these sermons amounted to little more than a gloss of sanctity on the proceedings. They also point to the continuing problem for contemporary astrologers of their precarious standing with the dominant religious orthodoxies. This legacy of the sixteenth century Reformation (which anticipates our period here) laid much of the intellectual groundwork for astrology's later disgrace, the social cement of which was provided by the events of 1642–60.

Despite carefully avoiding overt politics and cultivating religious legitimacy, the Society of Astrologers failed to survive the Restoration. Perhaps its members felt it unwise to call attention to themselves in the new milieu, as a collective group that had, far from fortuitously, sprung up and flourished during the Interregnum. That at least is what the other evidence would suggest. It was not altogether forgotten, though. By 1677, Ashmole had lost five friends, including Richard Saunders; and Lilly, to whom he felt closest, was aging. On 22 June of that year, he nostalgically 'summoned the remainder of our old Club about Strand Bridge [Lilly's old London home] that are left alive.' In 1682, there was actually a brief revival of the Society. Ashmole recorded in his diary, for 13 July, 'The Astrologers Feast restored by Mr. Moxon.'[86] A friend of Ashmole and Streete, Joseph Moxon (1627–?1700) was an instrument-maker, printer and hydrographer (the latter by appointment to Charles II). He was admitted as a Fellow of the Royal Society in 1678, but apparently expelled for arrears in 1682.[87] Lilly and Wharton had both died the previous year, along with many other former members; with new generations of astrologers present, the attendence at the restored Feast must have been quite different from the original.[88]

The second and last Feast of the restored Society of Astrologers was held on 29 January 1683, at the Three Cranes Tavern on Chancery Lane. (This inn had provided the former residence of Richard Saunders.) The Stewards were William Wagstaffe and Sir Edward Dering (*fl.*1660– 1701). Little is known about the former, except that he was a Town Clerk of the City of London at the time.[89] Dering, however, was quite an important person in the post-Restoration astrological community. Half-brother of a better-known Lord Commissioner of the same name, he was a London merchant and staunch Royalist, who had cheered Charles II in exile with predictions of the king's eventual triumph. He was introduced to Ashmole in 1681 by John Butler, the Oxford Non-juror divine and astrologer. Three years later, Gadbury dedicated a book to the 'constantly

Loyal (my ever Honour'd Friend) Sir Edward Dering'. More surprisingly, Partridge, a radical Whig astrologer, also claimed Dering as a patron; one of his books too carried a laudatory preface by the merchant courtier. (Perhaps Partridge was glad to have such support from a quarter whence it was otherwise very rarely forthcoming.) Dering's well-placed generosity and eceumenical dedication to astrology, like that of Ashmole, seems to have cut across other loyalties.[90] But that was no longer sufficient to keep alive that strange flower of the Interregnum, the Society of Astrologers and their Feasts.

3

Astrology in Crisis

After the Restoration

As in so many other respects, the Restoration of the Stuarts in 1660 was a turning-point in English astrology. Given the events of the preceding years, its effects – both immediate and in terms of a subsequent general milieu – were decisive. Most obviously but importantly, those who had fallen or fled after 1649, or their heirs, now returned to power led by the son of the 'martyr King'. The preceding two decades had vividly demonstrated that there was nothing ineluctable about their position, that life and even government were quite possible without the King, House of Lords, or bishops. Despite the fact that many among the general populace were also exhausted by the previous years' instability, the new rulers therefore felt their position to be insecure. New legislation in the 1660s, such as the Uniformity Act and the Clarendon Code, was introduced to strengthen the hand of both Church and State.

These developments not only had a direct effect in suppressing remaining radicalism (whether real or imagined); the climate of fear towards sectarian enthusiasm ensured that for a social programme of almost any sort to flourish, possibly even survive, it had to be able convincingly to distance itself from such associations. Fine distinctions between various kinds of subversive positions – whether popery, Hobbist philosophy, or animistic materialism – were rarely drawn, and the roots of all such putative threats to public order were invariably traced to the Interregnum. Furthermore, memories of that period – in which astrologers had so visibly played a role, even if arguably not a major one – stayed fresh for many years afterwards, revived by the unrest of the 1670s and 1680s.

As we have seen, judicial astrology flourished during the Civil War and Commonwealth, in the heady air of broken censorship. The many radical involvements of astrologers in those years, their prolific output and high public profile, meant that astrology subsequently became firmly identified not only with the Parliamentary revolution, culminating in regicide – Lilly was formally and closely questioned in 1660, about who had done the deed – but also with the 'Familistical-Levelling-Magical temper' of the radical sects.[1] Even the circumspect Society of Astrologers disbanded. There had long been critics, of course. The anonymous author of Lilies Banquet: or, the Star-Gazers Feast (1653) had perceived places set therein 'for all Sects and sorts of persons, both Presbyterians, Independents, Anabaptists, Quakers, Shakers, Seekers and Tearers.' (The material inaccuracy of this charge – Presbyterians were among Lilly's bitterest enemies, and of the others his only reliable allies were the Independents – is besides the point.) Such critics, formerly held in check, now had powerful new teeth and new recruits to their ranks.

The power of astrologers to influence and incite the people had also been clearly demonstrated. It was these two considerations, taken together, that were so damning. Unchecked by 'responsible' authorities, astrologers' almanacs – claiming heavenly and divine approval for irreligion and sedition – had sold by the thousand. Their textbooks, teaching such appalling skills as 'how to judge of the permanency or durability of Kings, or such as are in authority by any Revolution', had further spread the rot.[2] And their influence had been shown by events such as Black Monday, 29 March 1652 – the occasion recalled by John Evelyn, quoted above, when the concern raised by astrologers about a solar eclipse was such that 'hardly any would work, none would stir out of their houses, so ridiculously were they abused by knavish and ignorant star-gazers.'[3] Again, the fact that the day fell out fine and no awful events ensued – even some subsequent ridicule of the astrologers – is of secondary importance. The significance of the incident lay in its proof of the power of astrologers, uncontrolled by authority, to excite and potentially incite the people.

The new government moved quickly to end this intolerable situation. All books now had to be licensed by one of Charles II's secretaries of state, the Archbishop of Canterbury, the Bishop of London, or the Vice-Chancellor of Cambridge or Oxford Universities; the monopoly on almanacs was returned to the (carefully vetted) Company of Stationers and the universities; and only official newspapers were permitted.[4] The principal, indeed zealous instrument of change was the arch-Royalist and High Anglican, Roger L'Estrange. In 1662, he obtained a warrant to 'seize all seditious books and libels, and to apprehend the authors, and to bring them before the council.' Soon thereafter, he was appointed to act as 'Surveyor of the imprimery' (that is, printing presses), a position which he used to great effect aginst dissenters and radicals, including astrologers. He was rewarded with a knighthood in 1685.[5]

During these years, the contents of almanacs, Lilly's perhaps in particular, suffered heavy censorship. No matter how veiled, allusions to public disorder, disharmony between the Crown and the people, ecclesiastical rigidity, or imperfections of government were struck out. Intimations of celestial (and thereby divine) criticism were even more strictly pursued. When the *Book of the Prodigies, or Book of Wonders* (1662) appeared, for example – discussing fifty-four signs of imminent revolution and/or apocalypse seen in 1661, including blazing stars and double Moons – L'Estrange seized the copies, destroyed the presses, and imprisoned the printers.[6] The persons as well as the presses producing such material were at risk. L'Estrange remorselessly hounded the radical bookseller and publisher Giles Calvert (*fl.*1639–d.?1663). Calvert had published John Webster's criticism of the universities (and advocacy of astrology), Jacob Boehme, Winstanley's tracts, and much of the early Quaker, Leveller and Ranter literature. When he pressed ahead with more such material after the Restoration – essentially, a republican almanac – he was accused of 'instilling into the hearts of subjects a superstitious belief thereof and a dislike and hatred of his Majesty's person and government', arrested, and imprisoned. (Note the use of the word 'superstitious' here, so crucial in the history of astrology, as an all-purpose pejorative: at once political, religious and epistemological.) Released but soon imprisoned a second time, Calvert died in 1663 or 1664, probably in Newgate. When his wife Elizabeth (d.?1674) continued to sell books that offended the authorities, she too was twice imprisoned on L'Estrange's orders.[7]

The astrologer John Heydon was also imprisoned in 1663, and again in 1667, for treasonably calculating and considering the king's nativity.[8] An entry in the *Calendar of State Papers*, which may or may not refer to him, records the following examination in February 1667: 'Astrological predictions on the questions to Peter Heydon, astrologer of London, as to whether the fanatics shall have toleration in England, and whether the English may not be compelled to a neutral place for a treaty of peace.'

(Heydon's answers were both in the affirmative.) 'Though these answers are according to the rules of his art, he is committed to the Tower for them.'[9] The last sentence is fascinating. It simultaneously evinces the survival still of a quasi-official recognition, no doubt denied elsewhere but persisting here, that astrologers operated by meaningful rules, which could be obeyed or transgressed; and that probity in the former respect was, even so, no protection against the State.

Even coffee-houses and other places of public discussion were regarded as dangerous, and monitored by the government; in 1675, the former were briefly suppressed.[10] But the effects of the Restoration in these matters were not only directly repressive, as I have said. The indirect effects were perhaps its most powerful legacy, through the active formation of socio-intellectual structures – and in the longer run, attitudes – that could be regarded as fundamentally sound. The prime concern of the governing elite was to ensure its own stability and security, either through externally imposed discipline, or (preferably) through securing people's assent to their own governance. Notwithstanding internal differences, this concern was reflected everywhere in the post-Restoration evolution of English society, from monarchical and factional politics and the Anglican Church to the nascent Royal Society.

The Church's position after 1660 was signalled by the return of the bishops (although with the loss of church courts). It was strengthened by the Act of Uniformity in 1662, which so firmly rejected the comprehension or toleration of dissent. And that in turn was secured through the power of a new generation who, as H. G. Alexander puts it, 'had been brought up to equate Anglican beliefs with loyalty to the Crown'. (Although this loyalty was soon to be tested and partly transformed by the Crown's loyalty, in turn, to Catholicism.) At the same time, the Church returned to its practice of taking ministers almost solely from among graduates of Oxford and Cambridge Universities, where the *ancien régime* had been restored more completely than anywhere else.[11] Finally, we have noted the return of ecclesiastical censorship. It requires no great imagination to appreciate the extent to which these developments were deeply inimical for astrologers, now as hated by the old Laudians as they were deeply unwelcome – tainted friends, to be avoided by even the most liberal Latitudinarians. Nor were the direction and weight of the Anglican restoration against astrology contradicted by the occasional exception, such as an aging William Lilly accommodated as Church-warden of Hersham, Surrey, and granted permission (as a personal favour to Ashmole) to practise medicine by the Archbishop of Canterbury.[12]

Sensitivity to sectarian excesses, including astrology, was widely voiced in such tracts as *A Free and Impartial Censure of the Platonicke Philosophie* (1666), by Samuel Parker (1640–88). Parker was a Royalist High-Churchman who was Archdeacon of Canterbury and later (1686) Bishop of Oxford and FRS.[13] The target of his attack was not just 'Platonicke philosophie', which he described as 'an ungrounded and Fanatick Fancy'. He was particularly worried about the sect of 'Rosicrucians', led by none other than John Heydon (hence, 'Heydonian Philosophy'). Astrology was a central plank of Heydon's projected 'Society of the Rosie Cross', which drew on Florentine Platonist ideas. It envisaged mastery over nature, through the knowledge and manipulation of natural magical–astrological forces. With such ideas, said Parker, Heydon and his cohorts 'directly Poison mens minds, and dispose them to the wildest and most Enthusiasticke Fanaticisme . . . And what Pestilential Influences the Genius of Enthusiasme or opinionative zeal has upon the Public Peace, is so evident from Experience, that it need not be proved from Reason.' Significantly, Parker also described Heydon's ideas as 'contrary and malignant to true knowledge'.[14] Once again, we find religious and political considerations at the very heart of what we might otherwise consider to be a 'properly' epistemological question.

The Royalist scholar Meric Casaubon (1599–1671) described astrology as a species of profane divination, 'founded upon mere imaginary suppositions and poetical fictions, words and names which have no ground at all in nature.' Lest Casaubon be thereby enrolled into the ranks of prescient scientific moderns, it should be noted that he was also a determined critic of the Royal Society. The title of his book makes clear the nature of his concern: *Of Credulity and Incredulity, in things Natural, Civil, and Divine* (1668).[15]

Another attack came from an influential philosopher and divine in the Latitudinarian wing of the Anglican Church, Henry More (1614–86). As the leading Cambridge Platonist, More was naturally anxious to distance himself from enthusiasm. This was almost certainly the chief reason for devoting four chapters of his *Explanation of the Grand Mystery of Godliness* (1661) to 'a brief but solid Confutation of Judiciary Astrology.' In this work, More carried on the traditional religious criticisms of astrology by attributing correct predictions to 'Aiery Goblins, those Haters and Scorners of Mankind'. He reserved the right of legitimate prophecy for the biblical prophets, and restricted the influences (and therefore uses) of the planets and stars to light and heat. Such attempts to deprive astrologers of their disciplinary raison d'être, by apportioning their expertise between the Church on one hand and astronomers on the other, were hardly new. In the climate of Restoration England, however, they acquired a new force.[16]

More, Parker, and Casaubon were not alone. Another critique described astrological naturalists as dangerous 'enemies to Christianity'. Its author was Robert Boyle, the leading Restoration natural philosopher in England, and his concern was expressed in a privately circulated manuscript written in 1665 or 1666, entitled 'A Free Inquiry into the Vulgarly Received Notion of Nature'. As this implies, and in common with his contemporaries, Boyle was highly aware of the contingent social implications of different natural philosophies. This awareness is therefore the proper context, both intellectual and social, in which to consider (as we shall below) his attitude to astrology.[17]

That these men were all worried about astrology, despite quite serious differences in other respects[18] – is significant; it reveals the preoccupations held in common by the governing Restoration elite. That elite was not monolithic, however. For example, there was an important strategic difference between two major groups. Parker's prescription was typically High Church and authoritarian: 'The way then to prevent Controversies, and to avoid Schisms, is not to define [what is true or right], but silence groundless and dividing opinions.'[19] This effort was the thrust of L'Estrange's campaign. But Latitudinarians such as Boyle frowned on such direct and repressive discipline; in their view it was too liable to rebound. An important plank of their more subtle, hegemonic approach was the new Royal Society, which gained royal approval in 1662. Its leading Fellows were moderate ex-Parliamentarians and Royalists, who were also often leading Anglicans. (We shall return to this programme in a moment.)

As part of the same social process (immeasurably more than as part of a disembodied and inexplicable current of scepticism), it had also become increasingly fashionable to laugh at astrologers, at least in public. John Phillips made fun of Lilly and others in his play *Montelion* (1660). Samuel Butler (1612–80) found an appreciative audience, especially at court, for his satire *Hudibras*. In the first two parts, which appeared in 1662–3, Butler included a flagrant parody of Lilly as the astrologer Sidrophel.[20] Similarly, John Wilson's play *The Cheats*, presented in

1663, mocked the astrologer Mopus. Dryden's *An Evening's Love: or, The Mock Astrologer*, acted in 1668, confronted an astrological imposter with a real expert. And as late as 1695, the astrologer Foresight was presented as a muddled (if kindly) old fool, in Congreve's *Love for Love*.[21]

John Dryden (1631–1700), himself an astrologer, is an interesting figure in this context. In print, he mocked astrologers: 'Thus, Gallants, we like Lilly can foresee, / But if you ask us what our doom will be, / We by to morrow will our Fortune cast, / As he tells all things when the Year is past.' But as J. C. Eade has shown, the astrology in Dryden's plays was expertly handled. Furthermore, Dryden drew heavily upon astrology (suitably adapted) in writing his *Annus Mirabilis* (1667) in praise of the Restoration. And in private, he was a sincere student of the art, who had his own and a son's horoscopes cast by Ashmole. That of another son he set himself, observing in a letter that 'Towards the latter end of this month, Charles will begin to recover his health, according to his nativity, which, casting it myself, I am sure is true; and all things hitherto have happened to the very time that I predicted them.'[22]

Dryden thus conforms to the pattern we have already encountered among Royalist astrologers, or those at least with any degree of proximity to the powers-that-be, such as Wharton (a baronet), or Aubrey, Ashmole, or Dryden (all FRS). Astrology was for them a more-or-less private pursuit, and its publicly prominent exponents and 'illiterate professors' (usually ex-Parliamentarians) often fair game for their pens. Also in private, members of the gentry (up to and including Charles II) continued on occasion to consult astrologers such as Ashmole. As we shall see, even astrologers' critics (especially among natural philosophers) sometimes resorted to astrological reasoning. John Evelyn, the Royalist virtuoso whose scathing indictment of Lilly and his audience has already been quoted, mused cautiously on 12 December 1681: 'We have had of late several comets, which though I believe appear from natural causes, and of themselves operate not, yet I cannot despise them. They may be warnings from God, as they commonly are forerunners of his animadversions.'[23]

None of this should obscure the basic process at work, however. Set against the consequences of the restored monarchy, of the Church and ecclesiastical censorship, the work of L'Estrange, the official hostility of the Royal Society, the published polemics, the persistent suspicions attached to Lilly, and the travails of even Royalist astrologers like Heydon and Gadbury – against all this, the overall drift of events is crystal clear. That drift was firmly to marginalize and discredit astrologers as such (regardless of the individual astrologer's politics or personality); and thereby also to destroy the power of their astrology, whose natural audience was the popular readership of annual almanacs. On the desirability of this, and in their various efforts to bring it about, the different factions of post-Restoration England were united much more than they differed.

For an idea of the situation among judicial astrologers after the Restoration, it is instructive, as always, to consider William Lilly. Perhaps partly as a result of his censorship, and partly because of exhaustion, disillusionment, or fear on the part of erstwhile buyers, Lilly's annual sales fell from (reportedly) nearly 30,000, in the late 1650s, to about 8,000 in 1664.[24] In an event which reveals the official suspicion of astrology, and of Lilly in particular, he was also erroneously implicated in Rathbone's Plot of 1665. The government paper, *The London Gazette*, reported that the republican plotters had admitted picking 3 September for their attempt 'as being found by Lillies Almanack, and a Scheme erected for that purpose to be a lucky day, a Planet then ruling which Prognosticated the downfall of Monarchy'.[25] In fact, there was no such statement or suggestion in Lilly's 1665 almanac.

His popular reputation received a boost, if nothing to placate the authorities (in fact quite the reverse), with the terrible events that overtook London in 1665–66. First a serious plague spread through the city, killing at least 68,000 people. Then, on 2 September 1666, the Great Fire broke out. It burned out of control for four days, devastating the medieval city of London, including St Paul's Cathedral, the Guildhall, the Royal Exchange, and many Company Halls. Soon afterwards, it was recalled (which is itself interesting) that Lilly's *Monarchy or No Monarchy in England* (1651) had included a series of prophetic woodcuts. Among them was one showing corpses wrapped and shrouds, another showing a city in flames, and a third a fierce fire over which were suspended two twins. (Gemini, symbolized by twins, had long been associated by astrologers with London, being the sign supposedly ascending at the time of city's founding.) In late October, therefore, Lilly was interrogated by a Parliamentary committee. He 'admitted' having foreseen the plague and the fire, but under suspicion of involvement in starting it, directly or indirectly, he wisely denied not only any knowledge of that, but also of having known the exact year.[26]

Lilly was not the only astrologer to emerge after the Great Fire with an enhanced reputation. Richard Edlin (or Edlyn) (1631–77) had also issued the following remarkable predictions in 1664, based on the preceding year's conjunction of Jupiter and Saturn and other astrological considerations: 'me thinks we have too great cause to fear an approaching Plague, and that a very great one, ere the year 1665. be expired . . . there will be great Drought and Barrennesse. Conflagrations or great Destruction by Fire, during the effects of that conjunction, which will continue till the latter end of the year 1666.' Edlin hedged no bets, emphasizing in his Conclusion that 'I have in several places hinted, and have great cause to fear, do therefore once again premonish you of a great Plague in the year 1665. and pray God divert it.' (He also predicted for 1678 'uproars and disturbances about Religion in England, especially in London', as 'factious spirits do now endeavour to promote their ends and interest.' In 1678 the Popish Plot was in full swing.) This is almost the last we hear of Edlin, however. Like Booker and Streete, he was given employment as a clerk by Ashmole, which enabled him to survive into obscurity.[27]

For Lilly, this affair was his last major public appearance. Apart from the occasional tussle with L'Estrange over the contents of his almanacs, he lived quietly in Hersham, Surrey. He carried out his duties as the local churchwarden of St Mary's Parish Church, Walton-on-Thames, and continued his astrological and medical practice. Apparently increasingly involved in the latter, he obtained a licence in 1670. In 1677, he adopted another astrologer, Henry Coley (1633–1707), as his heir and amanuensis. When he died of a stroke on 9 June 1681, his old friend Elias Ashmole was with him. His death was a heavy blow to Ashmole; the two men and their wives had grown very close in recent years, as the touching correspondence preserved among Ashmole's papers reveals. Semi-respectable at last, Lilly was buried in his churchyard. A marble gravestone (still standing) was paid for by Ashmole; the stone bears an elaborate inscription by George Smalridge, a future Bishop of Bristol. Coley carried on the *Merlini Anglici* almanac, but the old fire was gone. It was fitting that an elegy for Lilly, printed by Obadiah Blagrave, related his death to the comet of 1680–1, and mourned his loss thus: 'Our Prophet's gone . . . The Stars had so decreed, / As he of them, so they of him had need.'[28]

The 1670s and 1680s saw new developments which threatened the overall post-Restoration consensus. The strains of an ever more overtly pro-Catholic monarchy, amid the furor of the Popish Plot (1678–9) and a general crisis over the question of Charles's successor, led to virulent new divisions between 'Whigs' and Low-Churchmen, and 'Tories' and High-Churchmen. At the same time, the popular movement led by Shaftesbury revived memories of 1641–2. The early 1680s saw Tory supremacy established, and the successful suppression of any remaining radical Whigs. Even the Royal Society acquired a distinctly conservative Tory cast in the years 1679–88, as the strategy of comprehension of dissent and hegemonic use of natural philosophy became the concern of a Low Church minority alone. To the High Church Tories, whose hand was strengthened by the accession of James II in 1685 (as it was weakened by his deposition in 1688), that approach still smacked of disloyalty and 'democracy'. Their ascendency was short-lived, however; revulsion against James's Catholic and authoritarian ambitions, both popular and on the part of a significant part of the establishment, resulted in the Glorious Revolution of 1688–9. This event set the lasting foundations for that characteristically English blend of monarchical and aristocratic power on the one hand and mercantile and professional power on the other.

These events only strengthened the downward course of judicial astrology, begun two decades earlier. The turmoil of the late 1670s and 1680s re-stimulated earlier Restoration fears of instability and subversion, including the involvement of astrologers. That association was even given a new lease of life by the inflammatory popular polemics of Gadbury and especially Partridge, the latter in his almanacs lashing popery, clergy and king, and arguing that 'a commonwealth's the thing that kingdoms want'. It was no coincidence that More chose to republish his anti-astrological polemic of twenty years earlier – and Boyle to publish his for the first time – in 1681. Around 1685, scrutiny of the contents of almanacs under the direction of L'Estrange and William Sancroft, the Archbishop of Canterbury, redoubled.[29] This subsided after 1688–9. Regardless of who had the upper hand, however, no faction of the governing elite – whose real differences should not obscure their common ground – had any reason to love astrology. Its affinities with popular disorder and insurrection, from which all those in or near power stood to lose, were now too clear. For that reason, the process of its exclusion from the dominant culture continued.

Even when radical Whig discourse reappeared (for example, in John Toland's tracts), it did so as pantheistic and deistic materialism. Those concepts did not exclude astrology, but neither did they favour what was by now an irremediably discredited social form for anyone attempting to gain the ear, or influence the actions, of the upper and middle classes. (As J. G. A. Pocock has pointed out, republican heirs of the Puritans such as Toland were now emerging paradoxically as Deists and opponents of the prophetic tradition; while it was the Anglicans, foes of the apocalyptic sectarians, who carried on that tradition, albeit very much on their own terms. Neither approach, however, left much space for judicial astrology.)[30]

Lilly was not the only astrologer noticeably missing from the briefly revived Astrologers' Feasts, in 1682–3. As early as 1669, a poem by John Gadbury lamented 'the Death of so many Eminent Astrologers, and Mathematicians, and particularly . . . the Expiration of Mr Vincent Wing.' (The others were Neve, Fiske, Oughtred, Culpeper and Booker.) Richard Saunders, the astrologer-physician who had relieved Ashmole's stomach pains after a Feast, died in 1675.[31] At the same time, there were indications of a growing malaise among practizing judicial astrologers. Joseph Blagrave's *Astrological Practice of Physick* (1672) contained this significant complaint:

> I find that many being unsatisfied concerning the legality of my way of Cure, have refused to come or send unto me for help to cure their infirmities: and many of those who did come, came for the most part privately, fearing either loss of reputation or reproaches from their Neighbours, and other unsatisfied people; and also fearing that what I did, was either Diabolical, or by unlawful means.[32]

A few years later, the astrologer-divine John Butler addressed Ashmole in an Epistle Dedicatory to his defence of astrology in 1680: 'But do you hear the news from Alma Mater [Oxford University]? All Astrology must be banished, and that so, as it shall not so much as find a room in the imaginations of men! Then what shall become . . . of me? And of us all Astrologers? And do you, Sir, think to escape?'[33]

Of the new and remaining astrologers, some (like Coley) attempted to stay out of the public arena, and concentrate on study and private practice. As Blagrave's words imply, however, that was no protection against the changes taking place. Many increasingly pinned their hopes for a revival, or simply survival, on a sweeping reform of their art. Behind this ambition lay fears, which acquired a new urgency in the aftermath of the Restoration, that 'this Art is so contaminated and defiled in general, and the true Grounds so conspurcated, and intermingled with many frivolous Fables, that except the pure Quintessence be extracted from those faeculent dregs, this Science which passeth all other Humane Arts, as the Light of the Sun the Stars, is likely to perish.'[34] Understandably, astrologers tended to see the solution to their problems as something more-or-less under their control, and therefore remediable by putting their house in order. But the messy pluralism of English astrology in the 1640s and early 1650s had not been considered by many of them to be a problem; it was the new pressures after 1660 that made it appear so.

Of course, there were still astrological activists. Francis Moore (1657–?1714) maintained a busy practice in London – first at Lambeth, then at Southwark. In 1699, he also started an annual almanac which combined astrological and medical advice with Whig politics. The editorship of *Vox Stellarum* was assumed by a succession of others after his death, and in the eighteenth and early nineteenth centuries, it went from strength to strength. Although it still bears his name today, Moore's almanac otherwise retains little of its original controversial and populist impulse.[35]

Moore's peer, the astrological physician William Salmon (1644–1713), has also received modern recognition of a kind: he is still uncritically recorded by contemporary historians of medicine as an 'infamous quack'. This description is a direct ideological descendant of the outrage of the Royal College of Physicians, when Salmon dared to extend Culpeper's and Blagrave's mission. (James Younge's rancorous polemic against Salmon in 1699 was rewarded three years later with an FRCP.) Salmon's principal crime was his popular handbook of medicine, whereby the poor could treat themselves: *Synopsis Medicinae: or, A Compendium of Astrological, Galenical and Chymical Physick* (1671). By 1699, it had gone through four editions. In addition, he published a comprehensive herbal for medicinal use in weekly instalments, and an astrological textbook, *Horae Mathematicae* (1679).[36] Salmon was also in bad odour with the authorities for his radical Whig politics. Upon James II's accession, he prudently removed to Holland. Upon his return in 1689, he resumed his practice and issued occasional almanacs from a number of London addresses. By 1704 he was receiving over a thousand medical and astrological enquiries a year. After Salmon, the tradition he embodied of eminent professional astrological physicians seems to have diminished, at least in the metropolis.[37]

Around this time, a new generation of prominent astrologers were throwing themselves into public controversy, notably the Tory–Whig struggle. Significantly, however, the most prominent – John Gadbury and John Partridge – were also the leading exponents of a reformed astrology, directed by quite specific sets of reforming principles. In this, they were conspicuously different from Lilly, Booker or Ashmole. Gadbury's reformism in particular initiated a number of similar tracts, beginning with his own *Collectio Geniturarum* (1662). These included Goad's *Astro-Meteorologica* (1686), Partridge's *Opus Reformatum* (1693) and *Defectio Geniturarum* (1697), and Godson's *Astrologia Reformata* (1697). The circumstances for these efforts were certainly not of the astrologers' own choosing, even though they had a hand in bringing them about. But in trying to win back a place for astrology in elite culture, they were obliged to try to adapt to the new terrain, make use of the available resources. Thus, the majority of astrological reformers divided into those led either by Gadbury or by Partridge: a natural philosophical, or 'scientific', programme and a Ptolemaic one respectively. The former was conducted by Royalists or Tories and Anglicans or Catholics; the latter by radical Whig Nonconformists. The organizing principle of 'camps' or 'programmes' seems justified, since in both cases men worked deliberately and quite closely together, in pursuit of a particular kind of reform, and in opposition to the other group. These two contending efforts, in other words, were each ambivalently aligned on opposite sides of the principal fissure running through the dominant elite and its ideology in post-Restoration England. As this division reminds us, that ideology was not monolithic; but it was fast becoming so with respect to astrology, leaving it stranded on the 'wrong', that is, plebeian, side.

A Scientific Reform

One of the new phenomena appearing in the aftermath of the Restoration, which, like so many others, bore the indelible impress of the preceding years, was the Royal Society of London. The experiences of its founding Fellows, a combination of moderate ex-Parliamentarians and Royalists, interacted with their natural philosophical interests to produce a distinctive programme: the creation of a body of undisputable, because 'true', natural philosophical knowledge. This was intended to be a form of knowledge, as Peter Buck puts it, 'free from the distorting effects of controversy and conflict'. Such knowledge would result from following a set of procedures specifically designed to produce it (or in its own idiom, discover it) under the aegis of the experimental method. By virtue of being based on the public and shared observation of empirical and therefore 'objective' facts, knowledge and procedures could then be presented as deserving the support of all reasonable persons. This was no intellectual fancy; the circumstances of the Restoration lent an unprecedented edge to considerations of how to guarantee assent. Another dimension was that of religious legitimacy. In the intentions of its proposers, such knowledge (far from encouraging secularism or atheism, 'scientific' or otherwise) was construed as confirming God's superior and active role in the universe, thus undermining dangerous ideas of materialism (whether animistic or mechanistic), with their levelling implications.[38]

However, facts neither present themselves uninterpreted nor interpret themselves. As Corrigan and Sayer have observed, 'The key question – then as now – as Hobbes and others realized, was "Who is to interpret?"' In practice, the production of natural philosophical knowledge was crucially reliant on a small elite community, the active nucleus of the Royal Society and their supporters, whose overarching concern was theological and political as well as philosophical soundness. Their experimental interpretations of Nature as God's book were intended to undermine and replace the competing prophecies of illegitimate, unreliable, and (literally) unlicensed prophets whose utterances, it was felt, had been so destructive and divisive in the recent wars. In the words of Shapin and Schaffer (who have analysed this process in detail), 'just as [religious] toleration was to be limited in order to be secure, so experimental liberty was distinguished from individual antinomianism and from uncontrolled private judgement'.[39] Obviously, this programme was an ambitious one, and despite some successes, its attempted realization was lengthy, uneven, and incomplete. (One constant problem was the vulnerability of its supporters themselves to charges of unorthodoxy.) But it was influential, and had particular relevance for astrology.

As we have seen, astrologers comprised the principal independent prophets of the Civil War and Commonwealth, before the world turned upside down was righted. Unlike the direct suppression of High-Churchmen and extreme Royalists, the Latitudinarian natural philosophical programme held out to them at least some hope. Precisely its apparent reasonableness and moderation seemed to say, 'all you have to do is show that astrology too is experimentally true, and you may join the club'. Is it surprising that some leading astrologers viewed the Royal

Society as a lifeline to the future, and responded to the siren call? Those who did so constituted the first group of reformers, whom I have described (with some liberty) as 'scientific'.[40] It was already a common-place among many late-seventeenth-century astrologers that in the words of one, 'the influences or the stars are purely Natural, and directed by Natural Beams, or Aspects Geometrical.'[41] But after 1660, the definition of 'natural' was itself hotly contested. At the same time, astrologers felt it increasingly urgent actually to demonstrate the truth of that assertion, and thus put astrology on a proper natural philosophical footing. Only this, they felt, would cleanse astrology of the superstitious and magical dross which had brought it into such disrepute, and reveal its objectively true (as opposed to infinitely malleable) nature. In short, their goal was to win back the lost approbation of high culture.

The three principal reformers were Joshua Childrey, John Goad and John Gadbury. Only the last was a professional judicial astrologer; but Gadbury and Goad were close personal friends, while he and Childrey were at least acquainted, and all three were conscious of working towards the same goal.[42] They also shared a very similar social character – Royalist, later Tory, and Anglican or Catholic – and a close association with the Royal Society in the 1670s and 1680s. In this period, the original Latitudinarianism of the Royal Society had become increasingly conservative; one indication was its interest in political arithmetic, 'a distinctly Restoration doctrine, concerned . . . with supplementing the power of sovereign authority, and facilitating its exercise'.[43]

What was the position of the Royal Society specifically on judicial astrology? An early promise of hostility had appeared in 1654, when Seth Ward, with the encouragement of John Wilkins, replied to John Webster's critique of the universities. Ward and Wilkins were both future founding Fellows and bishops. Responding to Webster's advocacy of including astrology in the curriculum, Ward railed against 'that ridiculous cheat, made up of nonsense and contradictions, founded only upon the dishonesty of Impostors, and the frivolous curiosity of silly people.' Of course, it is possible to set against this the favourable sentiments of other Fellows, such as Ashmole and Aubrey (at the risk of being misled by various red herrings, concerning who was 'more important' or even 'more scientific', whether a balance of some sort obtained, and so on).[44] It should be noted, however, that the closest thing to a statement of official ideology by the Royal Society left no doubt concerning its position on astrology: it was opposed.

That statement was Thomas Sprat's *History of the Royal Society* (1667). A Fellow (from 1663) and future Bishop of Rochester, Sprat seems to have been a rather sycophantic careerist. That in no way diminishes the historical importance of his book, ordered by the Society's Council, supervized by John Wilkins, and approved by Sir Robert Moray (President 1660–2), Lord Brouncker (President 1662–7), and John Evelyn. Other evidence suggests that Sprat's views were also shared by at least Christopher Wren and Robert Boyle. Although Sprat's views were more programmatic than substantive, they are therefore worth quoting at length, and the resonances worth noting. Astrology, he wrote,

> is indeed a disgrace to the Reason, and honor of mankind, that every fantastical Humorist should presume to interpret all the secret Ordinances of Heven; and to expound the Times, and Seasons, and Fates of Empires, though he be never so ignorant of the very common Works of Nature, that lyes under his Feet. There can be nothing more injurious than this, to mens public, or privat peace. This withdraws our obedience, from the true Image of God the rightfull Soveraign, and makes us depend on the vain Images of his pow'r, which are fram'd by our own imaginations . . . This affects men with fears, doubts, irresolutions, and terrors. It is usually observ'd, that such presaging, and Prophetical Times, do commonly fore-run great destructions, and revolutions of human affairs. And that it should be so is natural enough, though the presages, and prodigies themselves did signify no such events. For this melancholy, this frightful, this Astrological Humor disarmes mens hearts, it breaks their courage; it confounds their Councils, it makes them help to bring such calamities on themselves . . . which they fondly imagin'd were inevitably threaten'd them from Heven.[45]

This statement is revealing, and unlikely to have caused any dissent among Sprat's colleagues. In itself, however, it is insufficient as a guide to the relations between astrology and institutionalized natural philosophy at the time. We need also to understand the more hidden currents of hope and promise that astrologers perceived. To begin with, the legitimizing authority of Francis Bacon – so tirelessly invoked by Sprat, Glanvill and Boyle – was itself deeply ambivalent on astrology. Bacon's explicit hope for an '*astrologia sana*', a sane astrology, gave heart to the reformers. More than that, it also supplied them with a methodology. 'For astrology,' as one reformer wrote, 'wants its History as much as any other part of Philosophie; It being the only Via Regia, to its Perfection; and all other wayes being but by-wayes.'[46]

The prize text was Bacon's *De Augmentis Scientiarum* (1623), which recorded his opinion that 'As for Astrology, it is so full of superstition, that scarce anything sound can be discovered in it. Notwithstanding, I would rather have it purified than altogether rejected . . . But for my part I admit Astrology as a part of Physic, and yet attribute to it no more than is allowed by reason and the evidence of things, all fictions and superstitions being set aside.' Among the latter category, Bacon included 'the doctrines of nativities, elections, inquiries; and the like frivolities, [which] have in my judgement for the most part nothing sure or solid.' He also rejected the astrological houses, and the use of astral magic in talismans or sigils. However, he accepted the efficacy of the aspects, the zodiacal signs, the planets' apogees and perigees and the speed of their motion, eclipses, and the fixed stars.

Once astrology had been reformed, Bacon believed it would be possible to use these celestial indicators to produce reliable predictions of the weather, plagues, and crops, as well as of 'seditions [and] schisms' – 'in short, of all commotions or greater revolutions of things, natural as well as civil.' Despite his disavowal of horoscopes and elections, he even raised the hope of extending astral prediction 'to events more special and perhaps singular, if after the general inclinations of such times and seasons have been ascertained, they be applied with a clear judgement, either physical or political.' 'Nor,' he remarked, 'are elections to be altogether rejected.' And (as if this was not enough) 'I will add one thing besides (wherein I shall certainly seem to take part with astrology, if it were reformed); which is, that I hold it for certain that the celestial bodies have in them certain other influences besides heat and light.'

The purified and restored astrology that Bacon advocated was to be brought about in four ways: 'future experiments' or testing predictions, 'past experiments' or comparing past events with celestial phenomena, astrological traditions, which 'should be carefully sifted', and the evidence of 'physical reasons'. Thus, astrology was declared fair game for the famous inductive methodology. In another work, his *Historia Ventorum* (1622), Bacon also recommended studying the effects of celestial phenomena on the winds – as a way to discover the natures of the planets, as much as to learn more about the weather. This choice was not original, of course; there already existed a long tradition of medieval astro-meteorological research. But the advice was influential, coming from Bacon himself.[47]

There were no other comparable models for the English astrological reformers. While mention could be made of Johannes Kepler, it is difficult to gauge the extent of his influence in England. Kepler had devoted two works to the problem of astrological reform: *De Certioribus Fundamentis Astrologiae* (1602) and *Tertius Interveniens* (1610). Like Bacon, Kepler argued both against uncritical acceptance of astrology and against 'throwing out the baby with the bathwater.' Decrying houses, signs, elections, horaries, and nativities (although he indulged in considerable analysis of his own), he too sought a reformed astrology based on the aspects. He added new aspects to the traditional ones, and he both advocated and practised astro-meteorological research to test and refine these aspects.[48]

It took the extraordinary circumstances of mid-seventeenth century England to transform these suggestions and aspirations into a relatively coherent and urgent programme for astrologers. Before turning to that, however, we must look at an influential intermediary figure. The most famous living English natural philosopher of the Restoration was Robert Boyle (1627–91). The youngest son of the wealthy Earl of Cork, Boyle was very influential in the founding of the Royal Society. But in his writings on astrology too, we find ambivalence. On the one hand, he counted among the 'enemies of Christianity' those men who, 'granting the truth of the historical part of the New Testament (which relates to miracles), have gone about to give an account of it by coelestial influences, or . . . such conceits, which have quite lost them, in my thoughts, the title of knowing naturalists.'[49] But Boyle refused to extend this criticism to a blanket condemnation of astrology. In his essay on 'Suspicions About Some Hidden Qualities in the Air' (1674), Boyle speculated that planetary emanations, touching and affecting the atmosphere, 'may operate after a very differing and affecting manner.' To discover more about 'Celestial and Aerial Magnets', discussed in the same essay, he recommended recording 'the moon's age, and her place in the Zodiac, and the principal aspects of the planets, and the other chief stars'. Boyle's unpublished notes also include two pages on 'Judiciary Astrology', in which he admits that 'though I cannot allow ye bold misapplications that Astrologers make of Coelestiall Influences, yet I am not willing to deny . . . that there are any such things as Coelestiall Influences, by which, I meane powers whereby Coelestiall Bodys may act upon sublunary ones, otherwise then by their Light, or their Heat.'[50]

But Boyle's fullest statement is contained in his essay, addressed to Samuel Hartlib, 'Of celestial Influences or Effluviums in the Air'. This document was probably written in the 1650s; but Boyle approved its

inclusion in his posthumously published *History of the Air* (1691). Here
he confessed that for him, the study of the planets would be impoverished
were it to be demonstrated that their 'mutual influences and virtues' had
no physical consequences nor uses, so that 'we know them only to know
them'. Clearly setting out his intermingled social and philosophical
concerns, Boyle continued that, although there are objections to the reality
of astrological effects,

> which are occasioned partly by the superstition and paganism incident to this
> kind of doctrine, partly by the imposture, ignorance, and want of learning,
> generally observed in the persons professing this kind of knowledge, partly
> by the manifest mistakes and uncertainty that there is in the predictions of
> this nature, and partly by the inexplicableness of the way or manner how
> they come to affect one another . . . it may, notwithstanding all those
> objections, still be certain, that these celestial bodies, (according to the
> angles they make upon one another, but especially with the sun or with the
> earth in our meridian, or with such and such other points in the heavens)
> may have a power to cause such and such motions, changes, and alterations
> . . . as the extremities of which shall at length be felt in every one of us . . .

He went on to recommend experiments: in particular, the keeping of
'an history or diary of the observations of the weather, and its changes in
all respects, and then an account of the several places, motions, or aspects,
each day, of the several bodies of the heavens, with the agreements,
doubts, or disagreements, that these bear one to another.' Almost
sheepishly, he remarked to Hartlib that 'You did not expect, I am sure,
I should have adventured into so particular an apology for astrology.'[51]
Boyle clearly thought there was a valuable kernel of truth buried in
judicial astrology, which he hoped to rescue from its vulgar and
superstitious exponents through research, and to explain in terms of his
corpuscularian theory. (The latter project seemed *prima facie* quite
plausible – the planets emitted 'effluvia' of a kind particular to each
different planet, and consisting of uniquely shaped particles; and the
medium of the atmosphere enabled these to reach and affect human beings
in their own ways, acting chemically upon their bodies.) The astrological
hopes raized by Bacon were thus nourished in the 1650s and 1660s by
Boyle, and they spread throughout his widespread circle of correspondents,
several of whom harboured similiar ambitions: Samuel Hartlib, John
Beale, William Petty, and Benjamin Worsley.[52] In 1657, even the
sceptical Christopher Wren – in his inaugural lecture as Gresham

Professor of Astronomy – declared that 'there is a true Astrology to be found by the enquiring Philosopher, which would be of admirable Use to Physick, though the Astrology vulgarly receiv'd, cannot but be thought extremely unreasonable and ridiculous'.[53]

With the work of Joshua Childrey (1625–70), we come to the more public and coordinated work that can properly be seen as a full-blooded attempt at reform. Childrey was educated at Magdalen College, Oxford. He kept a school from 1648 until the Restoration, when he became a beneficed clergyman. In 1663 he was installed as Archdeacon of Sarum, and he also obtained a position at Salisbury Cathedral, where Seth Ward, FRS and Savileian Professor of Astronomy, was the Bishop. Childrey was a lifelong orthodox Anglican. In philosophy, he was an equally staunch Baconian, relating in a letter to the Secretary of the Royal Society, Henry Oldenburg, that 'I first fell in love with Lord Bacon's philosophy in ye yeare 1646'. That love never wavered, although he combined it with a deep conviction of the truth of heliocentrism. He was also a competent astronomer, and well known to the Royal Society, particularly in relation to its meteorological interests. In 1664, he was sent a thermometer of the kind recently developed by Robert Hooke. (Unfortunately, it broke en route on the Salisbury waggon.) In the *Philosophical Transactions* of 1670, Childrey capably held up his end of a debate with John Wallis, FRS, the Savilian Professor of Geometry, on tidal theory.[54]

Childrey's first book, and his most important statement of the Baconian astrological reform, was *Indago Astrologica: or, A brief and modest Inquiry into some Principal Points of Astrology* (1652). In it, he described the need, and in his opinion the means, to provide astrology with a 'Foundation and Principle demonstrable'. The former was plain, he wrote, in both the weakness and the shifts of 'the old astrology'. In Childrey's view, its parlous condition called into question the traditional means, especially the five 'Ptolemaick aspects', used to make such predictions. By 'shifts', Childrey was referring to the body of complex astrological rules of interpretation – not only the signs and houses, but rulerships, dignities and debilities, elements, Arabian parts, and so on – without any apparent physical foundations, or even analogues. This confusing and contradictory body of doctrine had evolved, he felt, just in order to be able to account for the contrary outcomes of predicted phenomena.

Childrey's prescription consisted of rigorous heliocentrism – which he regarded as having been confirmed by Galileo and others – and equally thorough Baconian observation and induction. In his view, 'though Astronomy be corrected, yet Astrology (which judges mostly by the Aspects) remains yet uncorrected'. The planetary aspects ought to be taken not 'quoad nos', as they appear to us (geocentrically), but 'quoad naturam', as they are. Childrey therefore recalculated all the positions of the planets in heliocentric terms, carefully noting the differences this made in several cases. He also advocated replacing most of the old interpretive rules with astronomically sound judgements, based on 'things hitherto not much regarded': a planet's apogee and perigee, upper and lower culmination, rising and setting, and latitude. To this argument, Childrey sought to add 'another, although but a weaker, That the truth of this doctrine of Aspects is wonderfully confirmed by observation.' He testified (although unfortunately, none of his records survive) that he found events to match his heliocentrically based predictions, even when contradicted by the old aspects. The latter, by implication, must have misled astrologers, from antiquity to the present, into making countless false predictions. But it would be a serious misreading to see all this as a purely abstract set of considerations. In Childrey's view, it was just astrology's unreliability and plasticity that made it such a danger – both to its own future, and to society – in the hands of unscrupulous astrological publicists. In this connection, he attacked the apocalyptic tendencies of his astrological contemporaries, who had damaged the credit of their art by predicting an imminent Judgement Day from the solar eclipse of 1654: 'the World by their Logick,' he wrote, 'would have been by this time dissolved a hundred times and over.'[55]

Childrey next published the rather grandly titled *Syzygiasticon Instauratum* (1653), a heliocentric ephemeris (and the first such work published in English, along with one other that year). This included an outline of the observational and inductive part of his programme. Citing Bacon's advice, he urged the compilation of detailed lists of 'signall occurances' in the history of several counties, as well as all 'grand extremities of weather' in the past, in order to compare these events with the appearance of comets, new stars and eclipses, and the heliocentric positions and aspects of the planets.[56] Of course, such an immense task could only be accomplished through the cooperation of many people. Nonetheless, Childrey continued to search out and record natural phenomena. The final fruit of his efforts, *Britannia Baconia, or, The Natural Rarities of England, Scotland and Wales* (1660), restated Childrey's credo about the reform of astrology:

> There is much to be found out, if man did but well attend to observation, and doubt even the very Principles of Astrology, til they had examined the truth of them . . . The way to go forward in this excellent Art, is to look back and compare the accidents of men and States, with the influences of Heaven, and this will not only try the truth of the old Principles, but adde new ones: such (it is very likely) as the sons of [the] Art do not yet dream of.[57]

However, another programmatic statement at this point was not a promising sign; and there was little in the way of new observations, or precise suggestions for dealing with the mass of data that would have resulted from compiling such 'histories' (indeed, that did result in the work of John Goad).

Childrey's pioneering heliocentric ephemeris for 1653 was followed by another with the same title for the following year, by one Richard Fitzsmith.[58] Both these ephemerides were based on the work of another heliocentric (and Royalist) astrologer-astronomer, Vincent Wing's *Harmonicon Coeleste* (1651). And the first of them appeared in the same year as *A Double Ephemeris . . . Geocentrical and Heliocentrical* (1653) by Thomas Streete, a similiar figure and one whom Childrey knew personally. Childrey's thinking in this regard was thus not an isolated instance. The specific recommendations of his work were not always well received by judicial astrologers, however. And he seems to have received little encouragement from natural philosophers. Hooke's standardized form for recording weather observations was published in *The History of the Royal Society* (1667). The following year, Childrey convinced a certain friend to record the local weather for him, commenting in a letter to Oldenburg that this person 'is the fitter (I take it) for observations of this nature, because being not astrologically affected, or (rather) infected, he is more likely to be impartiall.' The following year, Childrey again wrote to Oldenburg about some recent aspects and the accompanying weather. He added, 'So that I doe not a little wonder (but I speake it under correction) why in that Synopsis of ye Weather for 2 or 3 days in Sprats history, ye Aspects of ye Planets should be omitted.'[59] Oldenburg's reply, if any, is unrecorded.

Childrey's reforms were far removed from any association with Interregnum radicalism. He was educated in Royalist Oxford; his ecclesiastical rise steady, if unspectacular, the direct result of the Restoration; and repudiation of any 'this-worldly' millenarianism. The rational and sound astrology he practised, and wished to bring about more widely, reflected just these values. That the methods he proposed for doing so were Baconian, in association with the Royal Society, points to the way the Interregnum radicalism with which Baconianism had been loosely associated had given way to Restoration conservatism. The cool reception he received nevertheless is a sign of contemporary educated suspicions. In Childrey's lifetime, there still seemed to be a chance to rescue judicial astrology from its vulgar and dangerous advocates; by the time of his death, the door had shut.

If Childrey made a brave start, the burden of the undertaking fell principally on John Goad (1616–89). After Merchant Taylor's School, Goad studied at Oxford University for his BA, MA, and BD During the siege of Oxford in 1646, he performed (in the words of Anthony à Wood) 'divine service under fire of the parliamentary cannon'. In the early 1660s, he twice delivered sermons at St Paul's Cathedral, and returned to Merchant Taylor's as the headmaster, where he remained for twenty years. However, during the Popish Plot of 1678–9 he was dismissed following accusations of Catholicism. In 1681, he moved to London and opened his own private school, and in 1686 – taking advantage of the milieu of James II's reign – he openly declared himself a Catholic. His close friends included Gadbury and Ashmole, and his work was known to the Royal Society.[60]

Goad's *opus magnum* was published in 1686 (the same year that Newton's *Principia Mathematica* was presented to the Royal Society). It was entitled *Astro-Meteorologica, or Aphorisms and Discourses of the Bodies Coelestial, their Natures and Influences, Discovered from the Variety of the Alterations of the Air . . . as to Heat or Cold, Frost, Snow, Hail, Fog, Rain, Wind . . . Collected from the Observation at leisure times, of above Thirty Years*. In 1690, a shorter edition appeared in Latin posthumously, edited by a former pupil of Goad's, entitled *Astro-Meteorologia Sana*.

Goad made a Herculean effort to substantiate the belief that the weather is influenced by the planets. At the same time, he tried to refine and reform astrology by recording and analysing weather–planet correlations. His work stands out in a long tradition of earlier attempts,[61] and contemporary ones by Childrey, Ashmole and Gadbury. In addition to trying to discover just what the most important influences were, his research was also an attempt to investigate and/or to substantiate some more general propositions. These can be summarized as follows: (1) The planets exert influence on terrestrial phenomena in a way that transcends their mere light and/or heat (although these may act as carriers). (2) The Earth's atmosphere – and hence the weather – is particularly subject to these influences, and therefore offers the best hope of discovering precisely what they are. (3) While the planets' positions in the ecliptic (i.e. the zodiac) and the diurnal circle are important, these considerations are neither as powerful nor as promizing as the planetary aspects. (4) By closely examining weather–aspect correlations, it will be possible to discern the specific nature and therefore influence of each planet, as well as its likely local and specific effects. This would not only yield more accurate weather prediction, but invaluable gains in prediction and control in other areas, such as agriculture and medicine.

Aspects are certain angular separations (usually with a small amount of variance permitted) between two or more planets in the ecliptic. It is a very old part of astrological doctrine that they have certain positive or negative effects. The reformers drew on this tradition to postulate that the properties of these aspects combine with those of the planets concerned to produce relatively precise effects on Earth, whether meteorological or civil. By implication, precise and reliable predictions (other things being equal) should also be possible. But aspects also possessed another crucial advantage: they were observable, and therefore (at least in principle) exactly calculable. By the same token, they were considered to be more likely to command general agreement than the non-observable and controversial parts of astrological doctrine, such as signs or houses, which lacked any clear physical basis or correlate.

Such desiderata are not self-evident. They resonate at a number of different levels, reflecting the fact that this programme was the complex product of various considerations. Simultaneously involved, but with increasing specificity, were the millenial astrological tradition, which supplied the fundamental ideas; the hopes raised by Bacon's work, which supplied the methodology; the Royal Society's more general socio-intellectual approach and its methods, which emphasized the desirability of empirically verifiable knowledge, in order to generate consensus and therefore stability; and the urgent need these astrologers perceived to reform astrology, in line with this approach, after its disastrous excesses in the 1640s and 1650s.

In 1652, Goad began keeping a weather diary, which he maintained until 1685.[62] In it he recorded non-quantitative observations of the temperature, precipitation (usually including the time of its onset), and wind (its direction and some idea of its force). He did possess various instruments, probably in experimental stages of development, and even succeeded in marketing his own barometer. But the great majority of his records were non-instrumental, like most of those at the time.[63] They typically read as follows – for example, over four days in November in the years 1655–58: (1) 'frost, faire, some winde'; (2) 'cloudy, rainy night'; (3) 'some raine morn, cloudy, wind'. (4) 'High windes.' (Such vagueness naturally combined with the inherent untidiness of weather conditions to contribute massively to the problem of evaluation.) In addition to this, from 1677 to 1689 Goad also sent monthly predictions of the weather, in advance, to Ashmole. These records were of the same form (and presented the same problems) as those in his diary. Even using only the five traditional aspects, there are ninety-three aspectual possibilities between the Sun, Moon, Mercury, Venus, Mars, Jupiter and Saturn. Goad was therefore obliged to choose specific configurations to test, such as Sun–Moon aspects, oppositions only of all the planets, and so on. Typically, he would list the weather on several days of the relevant aspect; sometimes he organized the information into tables for each aspect. He also compiled tables showing the totals for each of thirty kinds of weather conditions, and noted the accompanying planetary activity.

The difficulties he faced in using the data to test his hypotheses can be gauged from these comments – probably addressed to Ashmole – on the configuration of the Sun in opposition to the Pleiades, for each of its annual occurrences from 1652 through 1658:

1652 'Here you see Clouds, Windy Wet. That's once.'

1653 'Concordat cum priori.'

1654 'Grant it to fail; yet that something of moysture, and a sharpe wind, is a spice of the Pleiades, and without much beging may be yielded.'

1655 'This is to the tune of our Pleiades.'

1656 'Suspicious in some places. What think ye?'

1657 'Does it fail or no? 'Tis fair, I grant, but 'tis a high Wind, not without Hail, compare that with the sharp Wind above, and do not overlook the Wet that fell the fifth day.'

1658 'Now I am partiall, and think this is our Influences; but I must leave it to thine own Observation: 'tis good sometimes to believe our selves.'[64]

As late as 1689, Goad noted a particularly 'drenching, thundering day' in July, and remarked plaintively to Ashmole, 'I wonder on what account, for Jupiter trine Saturn will not justify it. Upon a quadrate of the Sun to Jupiter sometimes we hear [of] it in July, but sometimes will not make an Aphorisme.'[65] This passage points to the intellectual problems of Goad's programme, arising from the fragmentary nature of the data and the marginal nature of the effects (if any). For Goad and his contemporaries at their historical juncture, this situation presented special difficulties. Concerning unseasonable weather, Goad wrote in *Astro-Meteorologica*:

> It may be said again (as it is by some great men) in things of this Nature, that they are Casual. But the word (Chance) in Causes Natural, and determinate, speaks our Ignorance . . . For what is Uncertain and Confused, is Casual, and Casualty is inconsistent with Science, so inconsistent that it is not to be pleaded by any lovers of Learning.[66]

Goad was thus truly inspired by the Baconian ideal; and applied (as Bacon had suggested) to astrology, he hoped both to salvage what was of worth in it, and extend knowledge into new areas. He began, rather optimistically, by assuming complete causal determination of the weather, and therefore its amenability in principle to accurate 'prognostication': 'we acknowledge that we have made use of every Brise; for we, who do believe there is no Casualty in the leaste Puff, directly issuing, could do no less.'[67]

What methodological procedures attached to this admirably 'scientific' attitude? It is difficult not to be struck by the quandary of a little-known pioneer, struggling in conceptual terrain since subdued. Goad's usual procedure for judging the efficacy of an aspect was simply to see if its apparent effect occurred more than 50 per cent of the time; for example, rain fell on 140 days of a new moon, out of a total of 261 such days (where one-half would be 130.5 instances of rain). In his own words, 'To a Moyety therefore we are arrived in the days, and that is enough to prove the Aspect not to be indifferent; they are as Powers of Fifty, to the Motion of an Hundred. So 'tis an even Wager it Rains on One of the three days concerned'[68] – early days in probability theory. In any case, and of this Goad was well aware, there was the egregious problem of how generalized planetary influences manifest themselves specifically in different places. Indeed, the problem of local variation still haunts weather forecasters today. However, Goad was addressing the difficult question of what lay between causal and 'Casual', and using an appropriate language: that of epistemological wagers. His work confirms Hacking's point that in locating the origins of modern probability theory, we should not look to the 'high sciences' of astronomy, geometry, and mechanics so much as to 'those lowly empirics who had to dabble with opinion'. (It is equally true, as he adds, that 'In recounting the work of the empirics it is of no value sedately to say that they combine science and occultism, and then leave out the occultism.')[69]

Goad's conclusions – apart from the obvious and expected one that the planets influence the weather – were often in line with traditional astrology. Sometimes he differed, concluding (for example) that Jupiter as well as Saturn brought dryness, whereas Mars as well as the Moon brought moisture. What reception did such conclusions, and the procedures underpinning them, receive among his peers? The overall answer must be: ambivalent or undecided. Robert Hooke, the Royal Society's curator of experiments, refused to take him seriously, recording in his diary after a visit to Goad in 1675 that the latter had 'made Jupiter cold and malignant and many other whims about the weather'.[70] At a meeting of the Royal Society in January 1679, the President, Sir Joseph Williamson, enquired 'whether Dr. Goad had perfected his theory of predicting the quarter and strength of the Winds from Astronomy. To which Sir Jonas Moore answered, that Mr. Flamsteed had examined several of Dr. Goad's predictions, but had not found one of three true. But Mr. Henshaw had examined them continually for about two years about a month since, and had found not above one of four false.'[71] This is a curious exchange. Did Moore – a consistent opponent of astrology, despite his close friendship with George Wharton – misrepresent Flamsteed? For on 4 July, 1678, the Astronomer Royal had written to Richard Townley,

> To conclude my letter, I cause my man to transcribe you the remainder[?] of an Ephemeris of the weather for this month, composed by Dr. Goad, whose conjectures I find come much nearer truth than any I have hitherto met with. They are derived from the aspects and positions of the planets, and, to establish his rules, he has made thirty years' observations. I cause my man to note daily the accidents of weather with us and the station of my baroscope and thermometer here. I shall entreat you, when the month is over, to send me your notes . . . and I shall return you mine. You know I put no confidence in Astrology, yet dare I not wholly deny the inferences of the stars since they are too sensibly impressed on [us].[72]

Of course, it is possible that Flamsteed was left unsatisfied by his month's examination of Goad's predictions; but in that case, what weight

71

was this experience permitted, against either Goad's thirty years of observations or Flamsteed's earlier approbation of the truthfulness of Goad's conjectures? Or was this a case of unwillingness to express publicly a private interest? Whatever the reasons, it was typical that Moore's and/or Flamsteed's criticism was countered by the approval of another Fellow, Thomas Henshaw (soon to be Vice-President). The same lack of consensus was evident after the publication of Goad's book. William Molyneux, the optician and founder of the Dublin Philosophical Society, was hostile. In a letter to Robert Plot, he noted the occurrence of a great storm without any unusual accompanying aspects, and poured scorn on Goad and 'his foul Holes in Heaven'. But Plot, then on the Council of the Royal Society and Professor of Chemistry at the Ashmolean Museum, was a keen supporter of the astro-meteorological programme advocated by another Fellow, John Beale; and he was impressed by, and hopeful for, Goad's work.[73]

There was also a review of *Astro-Meteorologica* in *Acta Eruditorum*, the respected scholarly journal published in Leipzig. The review consisted mainly of a summary, but concluded that Goad's fervour in attempting to rebut Gassendi and other critics of astrology (whether reformed or otherwise) had led him into excesses of his own, such as hoping to deduce from planetary positions not only the daily weather, but also 'floods, comets, new stars, sunspots, typhoons, prodigious rainfalls and monstrous births'. Goad might have objected to this charge, having already written sardonically that 'As superstitious as we are, we don't undertake to reduce all Properties to the Visible Heavens'.[74] But his having recorded such information implies that his hopes did indeed lie in that direction.

The third man at the heart of this reform, and its most tireless advocate and coordinator, was John Gadbury (1628–1704). His life is fascinating, not only in its own eventful colour but for the way it reveals the sweeping changes occurring during his lifetime. For Gadbury's astrology saw a dramatic transformation running exactly parallel to one in his political and religious beliefs. Born in Wheatley, Oxfordshire, by 1648 he was living in London, where he became a follower of the Levellers (and first met Lilly). He subsequently joined Abiezer Coppe's notorious Family of Love. In 1652, however, he settled in Oxford, and began to study astrology seriously; his tutor was the mathematician and astrologer, Nicholas Fiske. In 1656, the first of a flood of books, almanacs and ephemerides appeared. These were not limited to astrology, but extended to prodigies, comets, the new colony in the West Indies, and navigation.[75]

After Oxford, Gadbury returned to London and settled permanently in Brick Court, by Dean's Yard, Westminster. By 1660, strongly Royalist sympathies were becoming evident in his almanacs. As he later put it, 'the Coelestial Orbs disown all Anti-Monarchical, Disloyal and Rebellious Principles.' This move from sectarian to monarchist sundered Gadbury from the leading Parliamentarian astrologer Lilly, as it was later to do from the the Whig, Partridge. In fact, Gadbury moved so far to the 'right' that he was twice arrested on suspicion of involvement in (real or imagined) Catholic plots: in 1679, during the Popish Plot furor, when he was also burnt in effigy by a mob; and in 1690, following the Glorious Revolution. (Both times, he was released for want of evidence.) When Gadbury died in London in 1704, he was buried in the crypt of nearby St Margaret's Church, where he had attended services. By then he was a notorious Jacobite and crypto-Catholic, as well as a nationally known astrologer. Only Lilly before him, and Partridge after, achieved such eminence.[76]

Gadbury's friends and colleagues included many members of the Royal Society, of whom at least three – Jonas Moore, John Collins, John Aubrey and Elias Ashmole – were highly supportive of his efforts to reform astrology. Indeed, Aubrey was engaged in his own life-long investigation of the subject, perhaps hampered, as he put it, by 'having from his birth (till of late yeares) been labouring under a crowd of ill directions.' As he wrote in a note attached to the geniture of his friend, William Petty: 'Italian proverb – "E astrologia, ma non é Astrologo," i.e. we have not yet that science perfect; 'tis one of the desiderata. The way to make it perfect is to gett a supellex of true genitures; in order where unto I have with much care collected these ensuing, which the astrologers may rely on, for I have sett doune none on randome, or doubtfull, information, *but from their owne mouthes.'* This desire to 'perfect' astrology through the collection of accurate birth data – the interpretation of which could therefore be relied upon to be true, or else permit unequivocal correction – was a principal motive behind Aubrey's unpublished 'Collectio Geniturary', which in turn formed the basis for that acclaimed collection of biographies, *Brief Lives.* From 1669 to 1686, he assiduously collected the birth-times (usually from their subjects), and erected the nativities, of men such as Thomas Hobbes, Walter Charleton, Robert Hooke, Edmond Halley, Christopher Wren, Richard Burton, John Dryden, Matthew Hale and John Evelyn.[77]

Some other Fellows were more sceptical. Henry Oldenburg, for example, thought one of Gadbury's almanacs worth buying, but only for its 'natural observations.' Aubrey's close friend Robert Hooke thought Gadbury 'mad' and astrology 'vaine'. That did not prevent him from calling in to see Aubrey and Goad, as well as (less frequently) Lilly and Gadbury, or meeting them in the one of the coffee-houses that they all loved to frequent, such as Garraway's (in Change Alley, Cornhill) or Child's (in St Paul's Church Yard). In 1673, Hooke purchased 'Childreys new astrology', as well as (in 1675) Christopher Heydon's *Astrological Discourse*, almanacs by Streete and Childrey, and an old book by Lilly; and in November 1688, the month of William's invasion-by-invitation, he even perused Lilly's prophecies of 1655.[78]

On another occasion, Aubrey advised Edmond Halley, Flamsteed's successor as Astronomer Royal, to give astrology more serious considera-tion. (We now know, whether or not Aubrey did, that Halley had already studied and cited the work of some sixteenth century astrologers.)[79] This occurred during one of Gadbury's brushes with the powers-that-be. Halley's reply also illustrates well a number of points: the private credit of astrology, but the opprobrium attached to the label of 'astrologer'; the tenacity of that label even respecting astrologers with overtly natural philosophical interests and connections, like Gadbury; and the resulting and understandable reluctance of men such as Halley to be visibly associated with astrologers or their enterprises. Halley wrote to Aubrey:

> As to the advice you give me, to study Astrology, I profess it seems a very ill time for it, when the Arch-conjuror Gadbury is in some prospect in [*sic*] being hanged for it, however I went to the Library and lookte out the book you recommended to me [Cyprian Leowitz's *De coniunctionibus*] which I find to be published in 1557, so that I doubt not but the more moderne Astrologers having more experience of things may have added to him considerably, however upon your recommendation I will read it over.[80]

Like his political and religious views, Gadbury's views on astrology changed radically. The impetus for this change can clearly be seen in the social and intellectual pressures on judicial astrology in his lifetime. He started professional life as a thoroughly traditional judicial astrologer; his first book, *Genethlialogia, or the Doctrine of Nativities, and the Doctrine of Horary Questions* (1658), bore a glowing imprimatur from Lilly. Soon afterwards, however, he criticized Lilly for having done astrology a grave disservice. What he objected to was Lilly's publication of the popular prophecies attributed to Mother Shipton, Merlin and the Sybils – all 'fit only for laughter'. Gadbury wrote vigorously against the wilder reaches of astrology, arguing with ever increasing justification in 1666 that 'It is by reason of the apocryphal part of astrology that the sound part so extremely suffers.' At the same time, he began to advocate and engage in research on astrology, echoing Childrey that 'Astrology wanteth its history'. In this enterprise, 'One real experiment is of greater worth and more to be valued than one hundred pompous predictions' – that is, sweeping and apocalyptic predictions of the kind so popular among his contemporaries. Gadbury's reasonableness is striking. Between those who believe nothing of astrology and those who believe everything, he wrote, 'I know not which of the two doth Astrology the greatest injury. I conceive she is equally indebted to them both.'[81] It could easily, but superficially, be inferred that Gadbury wanted to introduce standards where none were. Just as in any discipline or craft, however, judicial astrologers had standards; although as elsewhere, they were often a bone of contention in one respect or another. What Gadbury wanted was to do was replace these strictly 'internal' criteria with those being developed by natural philosophers, in order to stem (as he saw it) astrology's degeneration. Nor was this a presciently modern desire; it was a response to his and his art's precise historical circumstances.

That is also true of Gadbury's adopted methods. His idea of natural philosophical experiments and research shows that this notion had not yet attained the relative definition it has today. Thus, several of his almanacs contained 'experiments' which consisted of detailed analyses of particular and individual horary problems, qualitatively examined. But he also worked hard to gather reliable data, and encouraged others to do so, for use in the Baconian reform of astrology. Realizing the enormity of this task, he called for better exchange of astrological information, and close cooperation between researchers.[82] In 1662, Gadbury published *Collectio Geniturarum*, the first English collection of the detailed nativities of well-known persons. As with the similiar efforts of Aubrey, this was intended to provide the beginnings of a reliable science of nativities. In his almanacs for 1664 and 1665, Gadbury also asked his readers to send in the birth data and chief 'accidents' of all children born on 4 and 5 of September 1664 – precisely in the manner of a longitudinal psychological study today. His goal was to ascertain what characteristics were shared by persons born at the same time, but otherwise unrelated – in other words, which characteristics were astrologically necessary, and which were merely contingent.

Like the other Baconian reformers – especially his lifelong friend, John Goad – Gadbury also put a great deal of hope and effort into astro-meteorological research. His records testify to the same intractability of subject-matter and inadequacy of approach that plagued Goad's work. Late in life, Gadbury admitted his quandary: 'I have been a daily observer of Aireal Variety for almost 35 Years, as the Noble Lord Bacon directs as necessary: And though I have met with several Similitudes of Verity . . . yet, I must freely own to have met with other Arguments too hard for me to bring under a Regimental Order of Experience.'

However, Gadbury's difficulties only increased his admiration for Goad and his *Astro-Meteorologia*, 'which Book is a most excellent Pattern for After-Astrologers: and of its kind, I may justly say, The World never saw its Equal.' This approbation contrasts painfully with Goad's more typical uncomprehending or dismissive reception in other quarters. By the end of his life, Gadbury himself was aware that his programme was approaching failure; indeed, he was almost its last surviving advocate. It is not surprising, therefore, to find these poignantly ambivalent reflections in his last almanac, for 1703:

> I am very much ready to part with any Errors, upon an assured Conviction they are such; yet, I shall not, cannot, wholly Renounce, or bid Good Night to Astrology: Lest in so doing I should Espouse a far greater Error, than any I am willing to part with. For Astrology is the language of the Heavens; and the Royal Psalmist says, The HEAVENS declare the GLORY of GOD. Howbeit, for my Great Creator's Honour, the Welfare of the Church and Nation, and Benefit of true Philosophy, I wish this Noble Art were well corrected.[83]

With the death of Gadbury, we can regard the scientific reform of astrology as moribund. Sympathetic Fellows of the Royal Society – such as Beale, Ashmole, Aubrey, and more carefully, Boyle – had themselves died by 1700, and in this respect they went unreplaced. It is true that the programme, and Gadbury in particular, had a successor of a sort: his colleague George Parker (1654–1743).[84] As we might by now expect, Parker was a Tory and High Church partisan, who became Partridge's chief rival after Gadbury. Parker edited and published Gadbury's last twenty-year ephemeris; he also produced an almanac containing detailed information on the planets' geocentric and heliocentric positions, accompanied by the more colourful and speculative elements that the majority of his readers demanded and paid for. It is significant that Parker concentrated on his almanacs, and made no serious attempt to carry on the reform programme. Nontheless, as a late member of the same group, he confirms its social and intellectual character. Parker's almanac for 1690 carried a commendation, for its astronomical accuracy, by Edmond Halley. It also appears that some of his data were also supplied by John Flamsteed (albeit without permission to say so).[85] In 1704, Parker took over the editing of *Eland's Tutor to Astrology*, earlier editions of which had been published and sold by Joseph Moxon, FRS. (Moxon had been responsible for briefly reviving the Society of Astrologers in 1682–3.)

There were a couple of other more isolated tracts, by astrologers about whom we now know virtually nothing, urging astrological reform along the same lines. William Hunt's *Demonstration of Astrology* (1696) sought to rescue the subject, now 'sullied over with Innovations and Superstitions', by showing it to be 'Grounded on the Fundamental Rules of the Copernican System and Philosophy.' Similarly, Robert Godson's *Astrologia Reformata* (1697), dedicated to 'the Royal Philosophical Society of London', deplored the 'utterly depraved and corrupted' state of astrology, and called for its 'Experimental Proof or Demonstration'.[86] These publications, with their stirring but ultimately empty calls to action, only emphasize the perceived urgency of reform among astrologers, while underlining its failure.

The social character of these reformers has been described. So too has the Baconian or 'scientific' approach, in association with the Royal Society, that they pursued. By now, the connections between the two should be plain. The acute social interests of the Royal Society in this period are unmistakable, and these reformers, on its fringes and with both friends and an intended audience within it, shared those values. With respect to consistency more than reductionism, it is not surprising that they also shared its intellectual and ideological commitments – notably Baconianism, the experimental philosophy, and a passion for quantification and correlation in areas where probabilistic induction (as distinct from experimental demonstration) was required.

As I have already pointed out, such an approach was intended, in the sensitive post-Interregnum climate, to provide a form of knowledge free from controversy and conflict. The production of such knowledge was a key task for Restoration natural philosophy, and the members of the

Royal Society (who were simultaneously leading members of the Anglican Church) were its prime movers. Together with the restored aristocracy, Church, and universities – often more conservative still – these aspects of the Restoration constituted the new milieu within which the reformers were trying to find a niche, and hence a new lease of life, for astrology. In their view, astrology too needed precisely such freeing from distortions, if it was to escape – indeed, if it was not to deserve – disgrace and discredit. As it happens, they failed to produce the empirical regularities that might have crowned their efforts with success. (I say might, because by then the weight of other considerations against astrology was so considerable.) Assuming they had succeeded, however, one thing is almost sure: the result would have been sharply distinguished, both by name and in substance, from its common and dangerous origins in judicial astrology.

A Ptolemaic Reform

The scientific reformers were not the only ones, however, and their claims to represent the only hope for astrology did not go unchallenged from within the astrological community. The challenge came from a smaller but at least equally vociferous group in the 1690s and very early 1700s, who were responding to the same crisis as the first, but with very different methods and values; where the inspiration of the former was Bacon, that of the latter was Ptolemy. Bacon was simply part of the problem: astrology's corruption and decline. And where the former were Tories and High Anglicans or Catholic, the latter were Whigs and Dissenters.

Much of the context for the first reform programme applies equally to the second. The overall problem was the same: the serious decline in the status of judicial astrology. So too was the root of the problem: the nature of its reputation after the Interregnum, especially among the upper classes and educated elite. Even the solution sought in this situation was broadly similiar, in so far as credit was to be restored (they hoped) through reform. Once astrology had been turned into a science that was reliably, and therefore demonstrably, true, it would naturally regain people's admiration and approbation – or rather, the approbation of those who were (in the opinion of these astrologers) qualified to judge these things, and to whom their appeals were addressed; that of 'the people' could be expected to follow.

In the decades following the Restoration, various crises ensured that the memories of Interregnum turmoil remained vivid. Indeed, the struggle for stability was just as urgent in the 1670s and 1680s as before, and political and religious heterodoxy as threatening to those in power. In 1675, Charles II clashed with Parliament in the Exclusion Crisis, over his attempted indulgence for Catholics (and Dissenters). The furor of the purported Popish Plot, amid widely accepted reports of subversion and sedition, took place in 1678–9. The accession of James II in 1685 was accompanied by the rebellion and brutal suppression of Monmouth and Argyll. Finally, intense uncertainty surrounded the Glorious Revolution of 1688–9 – an event of unwelcome proximity to the events of 1642–60, and one which raised uncomfortable questions about loyalty and resistance to monarchical authority. Furthermore, it is important to notice the high profile of astrologers such as Gadbury and Partridge, of the radical 'right' and 'left' respectively. The widespread dissemination of their views through almanacs was guaranteed to raise the ghosts of Lilly, Booker and Culpeper. At least one critic, James Younge, made the connection explicit, and pointed to their common aim of destroying 'monarchy, episcopacy and nobility.'[87] Gadbury was widely suspected of involvement in the Popish Plot, while at the other extreme Partridge deliberately invoked the republicanism and anti-clericalism of the Good Old Cause. William Salmon carried on Culpeper's project of taking physic out of the hands of the College of Physicians, and putting it into those of the common people.

This second reform programme reached its high point in the 1690s, just after the Whigs had assumed power in the Glorious Revolution. Radical Whigs such as Partridge now had high hopes, and not without reason, after the dark days of James II. For reasons and in ways we shall see, however, these hopes were dashed – including, by the first decade of the eighteenth century, this extraordinary attempt to revive English astrology in terms of Aristotelian natural philosophy.

John Partridge (1644–1715) was born in rural East Sheen, and one biographer described him as always remaining 'something of the country boy'.[88] He initially worked as a cobbler, but studied Latin and some Greek and Hebrew. His astrological tutor was one Dr Francis Wright (about whom we know nothing more). Partridge soon moved to London, settling first in Covent Garden, and later Salisbury Street, Blackfriars. He began issuing an almanac in 1678; from 1680, it appeared under the title of *Merlinus Liberatus* – a conscious echo, perhaps, of Lilly's *Merlinus Anglicus*.

Partridge's convictions were passionately Whig and Dissenter, and his popular almanac provided an ideal vehicle for their expression, often in highly colourful language with no holds barred when it came to opponents. The latter category included many of his astrological colleagues, especially reformist astrologers such as Gadbury, whose politics and astrology he detested equally. More ambitiously, however, it also embraced the Stuart monarchy and Anglican clergy; the same biographer also termed him, correctly, 'a great enemy of the Church'. He was prevented from publishing at the beginning of the Exclusion Crisis, and when James II acceded to the throne in 1685, Partridge decamped for Holland. But he continued indefatigably to issue prophecies, notably his *Annus Mirabilis* of 1688, which forecast the downfall of James and popery. With the accession of William and Mary, he returned joyfully to London. By this time, Partridge was almost certainly the most famous, or notorious, astrologer in England.[89]

His fortunes dipped after 1707, when he was mercilessly satirized by Jonathan Swift (in an episode to be discussed below). The upshot was that he became a laughing-stock in coffee-house circles in London and even on the Continent. At about the same time, he was embroiled in a dispute with the Company of Stationers which prevented the publication of his almanac for three years (1710–13). But there is no reason to think that either incident diminished his popularity, or indeed his sales, before or after this interruption. Late in life, Partridge left London and settled in Mortlake, Surrey, near his birthplace. When he died in 1715, he left behind £2,000 in his will.[90]

Turning to his astrology, Partridge was the acknowledged leader of this second and very different reform programme. Its substance was contained in his two principal books. The first, *Opus Reformatum*, appeared in 1693, and announced his intention of 'Reviving the True and Ancient Method laid down for our Direction by the Great Ptolemy.' He lamented all the cheats and frauds, especially Gadbury and 'his crowd of Errors', and described his own study of astrology as 'no otherways than a Branch of Natural Philosophy, and [I] do think it is no hard matter to give it a fair Foundation on very rational Principles, and those I think demonstrable too without any great Difficulty and Trouble, and they are Motion, Rays, and Influence.'[91] His second book, *Defectio Geniturarum* of 1697, was based on a critique of Gadbury's *Collectio Geniturarum*

(1662). In it he attacked Gadbury as an 'ignorant Reformer of Astrology', peddling both erroneous data and methods of interpretation. Also denounced were Gadbury's colleague George Parker, Kepler, a 'witty man, and an Enemy to Astrology', and heliocentric astrology in general. On the other hand – in marked contrast to Gadbury's old enemy of an earlier generation, Lilly – Partridge also asserted astrology's rationality, and condemned 'Magick-Mongers, Sigil-Merchants, Charm-Broakers, &c.'. Lilly would have appreciated the book's substantive content, however; it consisted of exhaustively detailed examinations of individual nativities. (Partridge spurned horary work as an Arabic invention, tainted with divination and lacking the warrant of Ptolemy.) In a deliberate contrast to Gadbury and his colleagues' importation of external criteria from Baconian natural philosophy, Partridge's astrology was puristically traditional. More generally, he restated his ambitions of 'Reviving and Proving the True Old Principles of Astrology.' There is a significant note of urgency in the second book, confirming the worries voiced by other astrologers: 'never did Astrology stand in need of a speedy Reformation more than at this time . . . Astrology is now like a dead Carkass.'[92]

What are we to make of this remarkable blend of apparently unrelated, if not incongruous, elements? The answer lies in Partridge's analysis of the problem, namely that astrology was in crisis due to cumulative effects of all the accretions and innovations in astrological doctrine since Ptolemy. Whether medieval, Arabic, or Baconian, they were all viewed as corruptions, which had only succeeded in bringing astrology into disrepute. This view implies his prescription: in direct opposition to the first group of reformers' modernistic and future-oriented outlook, his solution was to shed all such decadent developments by returning to the ancient purity of a pre-lapsarian past – in this case, an Aristotelian and specifically Ptolemaic purity. It is important to appreciate that from this point of view, astrology was a completely rational form of knowledge; but it was rational in the old Aristotelian sense. Partridge was therefore equally opposed to the popular magical astrology of (at its acme) Lilly and to the scientifically rational astrology of Gadbury. He clearly thought the latter represented a more dangerous threat, however – partly, no doubt, since he viewed it as an upstart corruption of true rationality, which had stolen its mantle. By implication, the naturalism of Partridge's natural philosophy can be contrasted with the Anglicanism of the natural philosophy he opposed. The implications of sectarian 'atheism', especially in the light of his other radical commitments, could scarcely have been lost on representatives of the latter. It was therefore with complete consistency that the text and touchstone for this reform was Ptolemy's *Tetrabiblos* – duly translated and published in 1701 by Partridge's principal colleague, John Whalley.

Whalley (1653–1724) was an almanac-writer and Whig polemicist. Although English, in 1682 he settled permanently in Dublin. (Dublin at this point was by far the largest town in the 'British' Isles after London, with a population of about 70,000.) He soon had a thriving business in astrological and medical consultation. Whalley's residence there was interrupted by his departure during its occupation by James II, after being publicaly pilloried in 1688 – a result of his diatribes against popery. Returning after James's defeat, he resumed production of his popular almanac, *Mercurius Hibernicus*. In 1714, he donned another cap, and began issuing *Whalley's News-Letter*, a non-astrological news and scandal-sheet. Whalley's success apparently inspired a pack of astrological imitators, competitors, and self-appointed successors in Dublin; his disputes with this unwelcome progeny lasted until his death.[93]

Whalley's substantive contribution to reform was his annotated translation, the first in English, of *Ptolemy's Quadripartite* (1701). This work, one historian tells us, 'created a great sensation at the time.'[94] It was dedicated to Partridge (along with Michael Cudmore, a physician in Drogheda), whom Whalley lauded in his text while lambasting 'the Adulterous Innovations' and 'voluminous Spurious Stuff' of Gadbury, Coley and Bishop. He warned that 'Young Astrologers from hence ought to take care what they read', avoiding those who were wilfully ignorant of the 'Truely Natural and Primitive Purity' of the Ptolemaic astrology, which Francis Wright had transmitted to Partridge alone. Its efficacy in interpeting nativities, particularly through the use of the 'Hyleg' and 'Anareta',[95] had been demonstrated in such 'Wonderful Prediction[s]' by Wright as the sudden 'Death of a certain Gentleman', which had baffled all other astrologers.

Whalley contrasted 'truly natural Astrology' with the doctrine of horary questions, which he called 'Imaginary, Un-natural, Arbitrary Whimsies, like those of Geomancy and the Common Astrology.' This attitude on his part confirms the remarks made earlier pointing to the naturalism and rationalism of the Ptolemaic astrology, opposed to both popular magical divination and Anglican experimental natural philosophy.

It may be noted, however, that in practice he was obliged by the scarcity of known birth-times to bend his principles a little. His almanac advertised that, among other services, he answered horary questions and drew up decumbitures. (Decumbitures are figures for foretelling the nature, duration, and outcome of illnesses from the supposed moment of their onset; they were a stock-in-trade of every early modern astrologer-physician.) The same was true of a rival astrologer in Dublin at this time, John Coats. In answer to Whalley's charges, Coats admitted answering horary questions, but added, 'I did not say (as he affirms I did) that it was according to Ptolemy's doctrine. Neither are Decumbitures, and I hope Mr. Whalley himself, will not disclaim the latter.'[96] Although obviously divinatory, Whalley could not wholly disclaim them, as Coats well knew.

Also in 1701, Whalley published a short essay entitled *An Appendix concerning [the] Part of Fortune, taken from the . . . Italian Astrology*. The 'Italian astrology' was that of this programme's curious third figure, *in absentia*: Placidus de Titis (or Placido Titi) (1603–68). He was an Italian Benedictine monk, described by Thorndike (aptly, if much too broadly) as the 'patron saint' of late-seventeenth-century astrologers.[97] In several books, particularly his *Tabulae Primi Mobilis* (1657), Placidus advanced the thesis that astrological influences are 'natural, manifest and measurable'. Each planet, in his view, had a unique nature which was manifest in its particular colour. The subtle influences of the planets were therefore contained in their light, and in that way transmitted directly to Earth. Like Partridge and Whalley, he too rejected 'interrogations'. Any similarity of rhetoric to that of some of the Baconian reformers, however, is misleading; Placidus's intent was Aristotelian, naturalistic, and rationalistic. As he boldly wrote, 'I desire no other guides but Ptolemy and reason.' It is not surprising that his ecclesiastical superior felt obliged to insert a reminder to readers of God's supreme status, over that of the planets, as First Cause.[98] Placidus also left his name on a new method of house division, based on sectors divided by measures of time rather than space. (The leading methods at this time were those of Regiomontanus and Campanus.) He claimed to have derived his system directly from Ptolemy's indications. Adopted and vigorously advocated by Partridge and Whalley, the value of this method was bitterly disputed by other English astrologers. They viewed it, correctly, as an emblem of Placido's larger claim – advanced by his English advocates – to have rediscovered 'the true and natural Astrology'.[99]

83

The last decade of the seventeenth century and first of the eighteenth saw a series of acrimonious exchanges between the leaders of each reform programme: Gadbury and Parker on the one hand, and Partridge and Whalley on the other. These attacks were both personal and theoretical – often simultaneously, as when Partridge accused Parker of whipping his wife 'the heliocentric way'. Parker naturally denied such charges, and in response to Whalley's imputation of 'Adulterous Innovations' assailed the idea that no progress had been made in knowledge since classical times. Why should Ptolemy be thought infallible in astrology, he asked, 'and yet so deficient in astronomy'?[100] A late contribution to the debate was *Flagellum Placidianum, or A Whip for Placidianism* (1711), by Richard Gibson (*fl.*1707–23). Gibson taunted the Ptolemaic reformers for practicing 'Monkish Astrology' (referring to Placidus), and accused them of trying 'to scourge us (the true Roman method) from our Heresie.'[101] After the Bickerstaff episode, however (see pp. 89–91), and now in his late sixties, Partridge seems to have lost some of his taste for polemic. Whalley, in Dublin, continued to issue occasional almanacs, but became increasingly involved with his non-astrological news-sheets. Not long after its rival natural philosophical rival, then, and despite their deep differences, the Ptolemaic reform too disappeared.[102]

The relationship between the social character of these reformers (including their politics) and their astrology raises certain questions. Why should Whigs, of all people, have been so apparently backward-looking, and – during the Whig, Latitudinarian and Newtonian ascendency from 1689 to 1702 – so anti-science? These puzzles, however, are not difficult to resolve. It was a central part of Whig ideology that there existed an Ancient Constitution, which justified the rulership of a king by the consent of the people (without, however, actually rendering the relationship contractual). As Pocock put it, 'a vitally important characteristic of the constitution was its antiquity, and to trace it to a very remote past was essential in order to establish it securely in the present'. Articulated principally by Sir Matthew Hale, this idea was bitterly contested by Tory scholars. The question was, naturally, particularly acute around the time of the Glorious Revolution, and for some time afterwards. Furthermore, the experience of Whigs as the 'Country' party had led to their being obsessed with the issue of 'Court' corruption.[103] Thus it seems highly plausible that for the Whig reformers, Ptolemy's *Tetrabiblos* was astrology's Ancient Constitution. Ptolemy was the warrant to condemn and sweep away all the popish-monarchical-tyrannical corruptions of astrology, brought about by astrologers who had strayed from the fundamental text, and reinstitute 'the true Primitive Astrology'. Such an attitude would have resonated with their uncompromising Protestantism, with its emphasis on another fundamental text, the Bible, and its hostility to magical superstitions (for example, horary and popular astrology) and interpretive or 'Innovative' priesthoods.

This anti-clericalism indicates another reason why, for Partridge and Whalley, the past may have appeared more attractive than the present, at least after the first flush of victory in 1689. Their political radicalism was a direct heir to that of the Interregnum sects, and intellectually their rationalistic naturalism pointed in the same general direction. But the rift between these Old, or Country, Whigs and the new ones, which accelerated under George I, was already well under way. Indeed, this programme on the part of Partridge and Whalley — who were clearly to the 'left' of the Latitudinarian and Junto Whigs who were actually in, or close to, power — was so far backward-looking as to be radical, in the original sense of the word: a return to a former pristine state. (One is reminded here of another, and almost contemporary, advocacy of a return to Aristotelian purity, perhaps impelled by radical motives: that of Henry Stubbe, in the early 1670s.)[104]

At any rate, these reformers had more in common with the vocal but powerless Stubbe than with one of the heroes of the Latitudinarian Whigs, Newton; which brings me to the second puzzle mentioned above, their hostility to the dominant natural philosophy. But in this connection, one must again recall Partridge's and Whalley's standing as spokesmen for radical Whig and dissenting convictions with Interregnum antecedents. (These considerations were neatly combined in Partridge's first book, when he recalled with pleasure one of Cromwell's victories which had included the capture of nine parsons.)[105] Given such a radical and marginal position, both socially and intellectually, it is hardly surprising that they should have found Newtonian natural philosophy, let alone the still more conservative Royalist strand, highly unattractive.

Using the Ptolemaic reformers, it is also possible to perceive other changes in this period. Not only was there a rift between the oppositional Whigs before the Glorious Revolution and their successors afterwards. The differences between, say, Lilly in the Interregnum and early Restoration and Partridge in the 1680s and 1690s, enable us to compare old Dissent with new, and observe a striking change. The struggle between Partridge, perceived as Lilly's heir, and their mutual opponent Gadbury centred on contested versions of rationality – an ideal to which they both subscribed. It was apparently no longer possible for a leading astrologer to openly practise and advocate an explicitly magical astrology, such as that of Lilly. Events in the years since the Restoration had succeeded in separating naturalism, whether materialistic or Anglican, and magical illuminationism. The latter was now confined to a purely popular level, to which few persons of any higher learning or ambitions would willingly be seen to stoop.

As a concomitant of this process, Partridge and Whalley, while still of the 'left', had abandoned the populist–democratic epistemology of Lilly and earlier radicals for an alternative elitism which paralleled that of their opponents. Astrology was to be the province of either a Baconian natural philosophical elite (the inheritors of Ashmole's old magical elite), or a Ptolemaic natural philosophical elite. But neither had much time for the common or vulgar people. Even through this wayward prism, the division between post-Restoration popular (or rather, now, plebeian) and elite cultures, and the hegemonic dominance of the latter, is clear.

Before leaving the subject, it may reasonably be asked whether there were any reformers who do not fit the schema of Tory Baconians and Whig Ptolemians. It would be extraordinary, of course, if there were none. The most obvious is Samuel Jeake (1652–99), whose astrological diary was recently edited by Hunter and Gregory, and whose work provides the basis for this discussion.[106] Jeake was a merchant in Rye, Sussex. A Dissenter in religion, a moderate Whig in politics, and an assiduous autodidact, he is interesting in this context because of his lifelong attempt to investigate and reform astrology. His first work – typically, a detailed self-analysis – was in 1668, and a series of increasingly precise and quantitative tracts followed. Perhaps the most ambitious consisted of an attempt to derive astrological theorems from a comparison of the events of one year in his life (1687–8) with the solar revolution and its directions for that year. (A solar revolution is a figure for one year of life, based on the Sun's return to the position it occupied at the time of birth.) None of this work was published. Many of its details were influenced by the French astrologer and mathematician, Jean-Baptiste Morin (1583–1656). Morin's massive *Astrologia Gallica* was published posthumously, in Latin, in 1661.[107] From 1684, when Jeake first read it, Morin became his hero. That can be attributed, at least in part, to the somewhat idiosyncratic combination of ideas that both men already shared: hostility to Copernicanism, a passion for mathematical and quantitative precision, and the ambition to test and reform astrology. And of Jeake's sincerity in the last respect, there can be no doubt.

To some extent, then, Jeake stands as a genuine anomaly to the approach taken so far: a reformer whose background was Whig Dissent, but whose model was neither Ptolemy nor Bacon, but Morin. But the extent and nature of this exceptionality needs careful qualification. To begin with the obvious, he was not a Tory Ptolemian or a Whig Baconian. Perhaps this last point needs some elucidation; it is true that his approach was largely quantitative. In three respects, however – the omission of Bacon's name or authority, the exclusive concentration on the judicial as opposed to natural end of the astrological spectrum, and above all the emphasis on single-case studies – it differed significantly from the Baconianism of Gadbury and Goad. In fact, it bore at least as much similarity to the exhaustive analyses of individual nativities (including the emphasis on directions) of the Ptolemians. Jeake also differed in another respect: his relative isolation. In contrast to the other reformers, whose principles and methodology were shared not only among colleagues but more widely, through publications, Jeake was neither an instigator nor member of a reform programme. He was briefly in correspondence with Henry Coley and John Kendall, two astrologers who themselves bore a tangential relationship to the major programmes; and he was aware of some of John Bishop's and John Partridge's work. But these tenuous connections were never pursued, either personally or intellectually. Jeake's investigations seem to have been almost obsessively private and introverted, as if for him the crisis in astrology that demanded its reform was more of a personal question of belief than one which was also general and shared.

This isolation from other reformers seems to have been due partly to his being a provincial autodidact. His correspondence with the *Athenian Mercury*, a literary and popular science journal, points in the same direction; it rebuked him for querying Copernicanism by raising an argument for a stationary Earth.[108] The point is not whether Jeake was right or wrong; it is his distance from the leading metropolitan arbiters of opinion, both within and without astrology. We also need to look beyond the nominal categories of 'Whig' and 'Dissenter' shared by Jeake, on the one hand, and Partridge and Whalley on the other. The defiant radicalism of the last two men, in both respects, contrasts strikingly with the relatively comfortable niche occupied by Jeake. He seems rather to have embodied that stream of Dissent which, unlike the anti-trinitarian and deist kind, was in the early stages of acculturating to more dominant social values. (It was still sufficient to set one apart, however, as Jeake was well aware.)

These circumstances are all important in understanding why Jeake, a middle-class entrepreneur, evidently experienced little difficulty in reconciling his social identity with the pursuit of astrology, at a time when it was increasingly unpopular in scientific, professional, and literary circles, among people not themselves members of the aristocracy or gentry. The Baconian reformers sought to integrate (if not ingratiate) astrology into the elite values and ideas that the middle classes had adopted and emulated; the Ptolemians sought to use astrology in attacking those values, and proffering an alternative. What was happening to judicial astrology, to which their reforms were a response, was clear to both. Jeake, however, remained relatively insulated from the exigencies of this struggle, and from a sharp awareness of it.

I am not suggesting that with the failure of the two main reform programmes, astrology in England stopped dead: far from it. Even after the deaths of Lilly and Ashmole, astrologer-physicians such as Blagrave, Salmon and Moore were kept busy with clients and publications. I have also described the non-reform activities of Gadbury, Parker and Partridge. Besides the latter, other astrologers – if a diminished number – continued to engage in purely polemical prophecy. For example, John Holwell predicted the imminent collapse (in 1682) of non-Protestant Europe; he received a scathing rebuttal from the Catholic astrologer John Merrifield.[109] But there were other astrologers who held back from such kinds of exchanges, and quietly pursued their practices and publications. Henry Coley (1633–1707) was William Lilly's adopted son, and heir to his almanac and practice, both of which he carried on from his home in Baldwins Gardens, off Gray's Inn Road. He also wrote a widely used textbook, *Clavis Astrologiae elimata, or a Key to the Whole Art of Astrologie* (1669; enlarged 1676). Despite his apprenticeship to one of the masters of polemic, Coley seems to have steered fairly well clear of astrological, political and religious disputes alike – no mean feat. George Parker described Coley as 'a person of a quiet and peaceable disposition'. He was a friend of both Salmon and Ashmole, as well as other Fellows of the Royal Society such as Aubrey, Hoskins and probably Moxon.[110]

I am saying, however, that the world of judicial astrology was by now irredeemably fractured. As the reformers had feared, with their social and intellectual credit exhausted, astrologers' clientele was shrinking and slipping down the social scale. It is symptomatic that not even the relatively uncontroversial and respectable Coley could not escape this fate. John Aubrey recorded a revealing incident in a letter to Anthony à Wood, dated 21 October 1693: 'I forgot to tell you if I just called upon Mr. Coley as I was going out of Town; and he is very angry with you, because you term Astrologers *Conjurors*.'[111] Coley's sensitivity is understandable. Wood, as an educated intellectual, ought to have known better than use a vulgar misnomer; indeed, he must have known, thus deliberately employing a term of abuse. But by now, Wood's attitude was typical of his class. The failure of the reformers had left English judicial astrology exposed to humiliation without appeal.

The Strange Death of John Partridge

In 1707, Jonathan Swift engaged in a highly successful satirical assault on John Partridge. As a revealing as well as colourful episode, it deserves some consideration.[112] Tom Brown, another satirist and pamphleteer, had set a precedent with several earlier attacks on Partridge, including a parody of his prognostications. Appearing in 1685, 1690 and 1700, these were almost certainly read by Swift.[113] Posing as the astrologer Isaac Bickerstaff and keeping a very straight face, Swift's opening gambit was a mock almanac: '*Pre*dictions for the year 1708. Wherein the month and the day of the month are set down, and the Persons named and the Great Actions and events of next year particularly related as they will come to pass. Written to prevent the people of England from being further imposed upon by the vulgar Almanack ma*kers* (1708).'

Claiming to be in possession of a marvellous new technique, with which he intended to rescue the reputation of astrology (!), Bickerstaff sought to demonstrate its efficacy by making several detailed predictions. These included the deaths (in typical astrological fashion) of several eminent persons, including the French Cardinal de Noailles, and – 'upon the 29th of March next, about eleven at night, of a raging Feaver' – John Partridge.[114] A veteran of professional in-fighting, Partridge simply ignored this publication. It attracted considerable attention, however.

Translated and republished on the Continent as a serious prophetic tract, it even drew at least one published refutation.

On 30 March, however, there appeared *The Accomplishment of the first of Mr. Bickerstaff's Predictions, being an account of the death of Mr. Partridge the almanack-maker upon the 29th instant* (1708). This letter, purportedly by an anonymous member of the gentry, vividly portrayed the death-bed scene, including a lengthy and highly unflattering confession by Partridge. The author commended Bickerstaff for his remarkable prophecy, but noted critically – in a manner highly characteristic of the arcane controversies in judicial astrological circles – that he had erred by almost four hours, Partridge having expired 'about Five minutes after Seven'. Then Swift added his famous abusive 'Elegy on the Death of Mr. Partridge', beginning 'Here five feet deep lies on his back / A cobbler, starmonger and quack'.

Swift's second pamphlet, like his first, was quickly translated into several European languages, and word of the joke began to spread. One Dr Thomas Yalden, posing as Partridge, issued *Squire Bickerstaff detected; or, the Astrological Impostor convicted* (1709), in which he described his difficulties in combating the (putatively) general belief that he was dead. Partridge himself, in his almanac written in late 1708 (for 1709), now protested indignantly (if somewhat redundantly) that not only was he still alive, but he had not expired on the said 29 March. But in what appears to have been quite an independent dispute with the Company of Stationers, his almanacs after 1709 failed to appear for the next three years. In any case, Swift was ready for him. In his again pseudonymous *Vindication of Isaac Bickerstaff, Esq. against what is objected to him by Mr. Partridge in his Almanack for the present year 1709*, Swift insisted that the astrologer was indeed deceased. In the first place, Partridge's almanac was evidence that 'no Man alive ever writ such damn'd Stuff as this.' Secondly, the subsequent appearance of Partridge's almanac was nothing to the point, since those of Dove, Wing, Gadbury and others also continued to appear, despite the undeniable deaths of their namesakes. Furthermore, the only other denial of Bickerstaff's predictions had come from France, protesting that the Cardinal de Noailles still lived. Were we then also to believe the testimony of a Frenchman and Papist, as against that of Swift himself – a loyal English Protestant? (To be implicitly allied with the former must have been, for Partridge, the crowning insult.)[115]

By now, it seems, Swift's trick on Partridge, as well as its true author, was known in capitals all over Europe. If his intention was to make the astrologer a laughing-stock in educated and coffee-house circles, Swift had succeeded remarkably. Whether this had any immediate effect on Partridge's or judicial astrology's overall popularity, however, may be doubted. Any such effects, lower down the social scale, were long-term and indirect. There is no evidence that sales of Partridge's almanac, for example, suffered when it reappeared in 1713; and the Company of Stationers found it profitable to continue the imprint into the last decade of the century. This episode thus possesses a significance which makes it an appropriate way to see out the seventeenth century, and turn to the eighteenth. Partridge's 'death' aptly symbolizes that of English judicial astrology, as a respectable metropolitan pursuit. But it also points beyond itself, to what survived.

PART II

Life After Death: 1710–1800

4

Popular Astrology: Survival

Astrology in Popular Culture

The next three chapters concern English astrology in the eighteenth century, after what has been variously described as its disappearance, decline or survival.[1] The difference in these descriptions points to historians' confusion about what actually happened, which usually begins with an overly narrow or simplistic definition of astrology in the first place. In fact, all three descriptions are largely true, in different social and intellectual contexts. After the dust had settled from the upheavals of the late seventeenth century, it is important to appreciate both what had changed and what remained. Until its final decade, astrology in the eighteenth century remained comparatively stable. In seeking the fullest possible picture, then – and allowing for the complexities and variations that render them more of a spectrum than discrete categories – three basic astrologies emerge: 'low' or popular astrology, 'middling' or judicial astrology, and 'high' or cosmological–philosophical astrology. Although the last kind can be said to have disappeared, nowhere do we find the single or simple death of astrology. As Mrs Hester Thrale shrewdly remarked in 1790, 'Superstition is said to be driven out of the World – no such Thing, it is only driven out of Books and Talk.'[2]

The dramatic events of 1642–1710 centred on the judicial or horoscopic astrology of professional practitioners such as Lilly, Ashmole, Gadbury and Partridge. But as such astrologers were aware, theirs was not the only kind. As George Wharton complained in 1648, for the general populace 'Ptolemy may be something to eat for aught they know.' In 1679, Henry Coley wrote that 'the sole interest of country readers . . . [is] to be told the stages of the Moon, and the sign.' This remained as true in the eighteenth century as it had been in the seventeenth. The editor of *The Family Almanack* in 1752 was 'well satisfied that not one Man in a Thousand understands an Ephemeris, and not one in Ten Thousand . . . has occasion to make use of one.' For that reason the Wiltshire astrologer Henry Season, while lashing rural ignorance, decided in 1762 to omit the Moon's signs, and print simply the corresponding part of the body (that is, 'head' for Aries, and so on).[3]

Popular astrology comprised the astral beliefs and practices found mainly among rural and semi-literate labourers and yeomen, but also among urban artisan and working classes. As with any primarily oral and tacit tradition, direct evidence is rare. We are dependent mainly on surviving books and almanacs, the word of hostile critics, and (beginning in the early nineteenth century) the research of folklorists, engaged in the contemporary 'discovery' of popular culture.[4] Historical evidence is rarely complete, however, and as Roger Chartier put it, 'the ways in which an individual or a group appropriates an intellectual theme or a cultural form are more important than the statistical distribution of that theme or form'.[5] In this case, there is enough for us to say that popular astrology persisted, almost unchanged in its content, from about the late Middle Ages until at least the mid-nineteenth-century, probably throughout the British Isles.[6] As such, in the eighteenth century it remained an important part of the cultural life of the relatively poor and powerless. It is also possible (indeed, necessary) to go further, and get a sense of how astrology informed their perceptions and offered guidance in matters of vital concern – agriculture, husbandry, physic, and love-life – thus comprising a kind of plebeian science of life.

At this time, roughly 80 per cent of the English population still lived in the countryside or small towns, and was employed in agriculture or the processing of agriculturally produced materials. Only in the last quarter of the century did the drift to the cities and early industrial capitalism begin to alter that pattern.[7] The exception was London. Its large population, including a high proportion of gentry, professional classes and tradesmen, and its role as both cultural and political centre, stand out impressively; but in terms of England as a whole, London was just that: exceptional. For the vast majority, life was still lived close to the land, with its cyclic round of the seasons and the weather's constant vagaries, the inseparable cycles of planting and harvest, and the accompanying rituals, ceremonies, and festivals – their Christian overlay, and partial penetration, notwithstanding. Such people lived in the constant company of the Sun, Moon, and stars, whose influences and uses – starting with the seasons, tides, and menstruation, and extending to astrological kinds of divination – were self-evident. Indeed, most rural dwellers would not have needed even an almanac to follow the more basic astrological events. Weather permitting, the Moon's phases, conjunctions of the planets, and eclipses were clearly visible, as they are today. Nor was all this a solely rural concern; in the towns too, there was enormous popular interest in the weather, eclipses and comets.[8] (Street lighting with public oil lamps was initially a London phenomenon, gradually installed there between 1690 and as late as 1740.)[9]

Exceptional celestial wonders, such as comets, were always a matter of special concern, usually in relation to contemporary natural, civil or political events. In daily life, popular astrology comprised a set of crude but partly for that very reason extraordinarily enduring concepts of celestial meaning, centred on the Moon (in particular its phases) and extending to the Sun (its eclipses and the dates of equinox and solstice). At the 'higher' end, the zodiacal signs occupied by the Sun or Moon were also taken into account, but that was about the limit of the astronomical technicalities observed. This alone does not imply impoverishment, however; for these considerations found a remarkably wide application. In all of them, as Samuel Johnson correctly observed, the Moon had 'great influence in vulgar philosophy'.[10] Not surprisingly, as we shall see, this was viewed in higher quarters as an alarming sign of the survival of pagan idolatry. More accurately, if no more reassuringly, it signified popular syncretism, as in a contemporary popular benediction upon first sighting the new Moon: 'It's a fine Moon, God bless her.'[11]

In *Lark Rise to Candleford*, Flora Thompson recorded a pig-killing during her childhood in the 1880s, in rural Oxfordshire. 'When the pig was fattened . . . the date of execution had to be decided upon. It had to take place during the first two quarters of the Moon; for, if the pig was killed when the Moon was waning the bacon would shrink in cooking, and they wanted it to "plimp up".'[12] This was no local or short-lived curiosity. The same belief was held in Sussex and Suffolk, and existed in print at least as early as the *Husbandman's Practice* (1664), which advised its readers to 'Kill fat swine for bacon (the better to keep their fat in boiling) about the full Moon.' The same book also prescribed shearing sheep during the Moon's increase, felling timber from full to new

Moons, and generally – in relation to the Moon's phase and sign – when to sow seeds, graft, plant, take pills, purge, and cut hair.[13] *Thomas Tusser*, a virtually identical handbook, was first published in 1557. After four editions in the sixteenth century and three in the seveteenth, it reappeared in 1710 (and again in 1812) as *Tusser Redivivus*. Cast in rough-hewn couplets, Tusser was replete with lunar lore: 'From Moon being changed, till past her prime / for grassing and cropping, is a very good time . . . The Moon in the wane, gather fruit for to last / but winter fruit gather when Michel is past . . . Sow peas and beans in the wane of the Moon/ Who soweth them sooner, he soweth too soon.'[14]

There were many regional variations, most of them noted by early-nineteenth-century folklorists. A visitor to well-to-do friends in Yorkshire went unserved because the servants were out, 'hailing the first new Moon of the new year.'[15] It was well-known in the northern counties generally that 'A Saturday's change, and a Sunday's prime, was never a good Moon in no man's time.' But the same held true (with slightly different wording) in Norfolk.[16] In Devonshire, apples were said to 'shrump up' if picked when the Moon is waning.[17] In West Sussex, it was lucky to first see the new Moon over your right shoulder, but unlucky through glass, and the new Moon of May had the power to cure scrofula.[18] In Sussex, Berkshire and Devonshire, praying to the new Moon was believed to reveal to a maiden her future husband, in a dream the following night: 'New Moon, new Moon, I hail thee! by all the virtue in thy body. Grant this night that I may see, he who my true love is to be.'[19] In Essex, people picked mushrooms at the full Moon; in Cornwall, medicinal plants.[20] In Staffordshire, mothers of children with 'chin-cough' prayed to the new Moon for a cure, adding 'In the name of the Father, Son and Holy Ghost, Amen.'[21] Although it exceeds our brief here, Brand noted that 'among the common people in Scotland in the eighteenth century, the Moon in the increase, full growth, and in her wane, were the emblems of a rising, flourishing, and declining fortune.' Orkney Islanders objected to marrying or moving during its wane, and a halo round about it presaged rain.[22]

Much lunar wisdom involved the weather. In Yorkshire, a halo round the Moon presaged imminent change. From Devon to Lincolnshire, a bright star or planet 'dogging' the Moon meant wild weather. For the Cornish, 'A fog and a small Moon, bring an easterly wind soon'. In 1828, Robert Southey visited Keswick, in the Lake District. Writing to a friend, he recounted:

Poor Littledale has this day explained the cause of our late rains, which have prevailed for the last six weeks, by a theory which will probably be as new to you as it is to me. 'I have observed,' he says, 'that, when the Moon is turned upward, we have fine weather after it; but if it is turned down, it holds no water, like a bason, you know, and then down it all comes.' There, Grosvenor, it will be a long time before the march of intellect shall produce a theory as original as this, which I find, upon inquiry, to be the popular opinion here.[23]

Unusual weather stimulated the demand for astrological explanation, as with *Mr Partridge's Judgement and Opinion of this Frost. Foretelling how long it is likely to continue, according to the Rules of Astrology* (1709). Conversely, unusual cosmic events raised expectations of corresponding local phenomena; hence, William Beetenson's opportunity to produce *Prodromus Astrologicus: Being an Astrological Discourse, on the Effects of the Eclipse of the Sun (or Earth), on Monday, May 11, 1724.*

Another major concern was astrological physic (or medicine). *Aristotle's Book of Problems* and *Aristotle's Masterpiece* were a popular publishing phenomenon. In at least three versions and well over twenty editions in the eighteenth century alone, they advised readers on 'Divers, Questions and Answers Touching the State of Man's Body', so that they could be healed or merely informed 'without applying themselves to a Physician.' According to this Stagirite, 'The fittest time for the Procreation of male Children, is when the Sun is in Leo, and Moon in Virgo, Scorpio, or Sagittarius . . . To beget a Female, the best Time is when the Moon is in its Wane, in Libra, or Aquarius.'[24] Nearly as successful was Culpeper's *The English Physician Enlarged. Being an Astrologo-Physical Discussion of the Vulgar Herbs of this Nation.* Originally published in 1652, in the eighteenth century it saw twenty-one London editions (plus three elsewhere). This astrological botany, relating each herb to a planet or sign and hence a part or organ of the body, was a direct descendent via Culpeper of Paracelsus's sixteenth century populist and magical physic.[25]

Judging by sales, the demand was great. So was the opposition from elite and monopolist medical authorities. Culpeper's leading successor (whom we have already encountered) was William Salmon. Salmon had a busy practice; his *London Almanack* of 1704 notes that he received 1,500–1,600 letters a year, asking for personal or medical advice. He warned correspondents that they must pay their own return postage or receive no answer, for the total 'makes a considerable Summ'.[26] When he published a popular handbook, the *Compendium of Astrological, Galenical, and Chymical Physick* (with four editions between 1671 and 1699), Salmon received the same treatment as Culpeper had three decades earlier. At the turn of the century, seeking to win the favour of the College of Physicians and Society of Surgeons, James Younge dedicated to the College a swingeing attack on astrologers, and Salmon in particular. It was entitled *Sidrophel Vapulans: or, The Quack-Astrologer Toss'd in a Blanket* (1699). Younge pilloried such 'Impudent, ignorant Quacks and Empiricks' as Salmon for daring to practise medicine, and their audience 'the Mobb (few else read or regard his Gallymawfry)'.

Salmon also issued a semi-annual almanac, which drew Younge's fire; for as the latter wrote, going to the heart of the matter, 'Almanacks are Oracles to the Vulgar.'[27] But Salmon, who died in 1713, was among the last generation of nationally-known astrologers and almanac-writers; his colleagues Francis Moore and John Partridge died in c.1714 and 1715, respectively. Despite the loss of these figures, however, almanacs remained probably the principal means of disseminating popular astrology in the eighteenth century. Nearly all those started in the seventeenth century continued after their founders' deaths, usually under the same names, but edited by lesser-known figures. Quantitatively, individual circulation figures mostly held steady or declined. By 1760, the total sales of fifteen almanacs – most, but not all, astrological in content – stood at 476,000. This figure implies a readership of several times more – in a national population (during the same years) of between 6.2 and 7.7 millions, not an insignificant portion of the total. On the other hand, it cannot be described as a unqualified success. The national population doubled in the course of the century (although the real increase came only after 1760); illiteracy generally declined, while total active readership more than doubled; and the circulation of newspapers was by now considerably higher. In contrast, by the last decade of the eighteenth century, sales of almanacs had only risen slightly, to just over half a million, and the number of titles sold nationally had declined to nine.

Three were non-astrological, including a popular purveyor of earthy cynicism, *Poor Robin*. This last, sold by the Company of Stationers from 1662, averaged sales of 10,000 copies a year throughout the eighteenth century. It should be noted that *Poor Robin*'s scepticism included bitingly sarcastic parodies of astrology, and attacks on individual astrologers. Its existence points to a strand of popular thought that rejected astrology, along with virtually every other arguably serious or systematic kind of thought.[28]

However, these figures conceal the spectacular rise of the most overtly astrological and prophetic almanac of them all: *Moore's Vox Stellarum*. Francis Moore was another astrologer-physician, with a busy practice in south London. He offered a wide range of services, including (to quote an advertisement) 'Judgement by Urine or the Astrological way, which is surest, without seeing the Patient', that is, diagnosis from a horary figure based on the moment a sample of the patient's urine is received.[29] Edited by Moore until his death in 1714, this almanac had begun life as simply one of the pack.[30] By 1738, however, it was outselling all its rivals, at 25,000 copies a year. By 1768, 107,000 were printed annually, or almost twice all the others combined. (Compare the annual circulation figure for what one historian has described as 'the most successful of all eighteenth century magazines' – that is, presumably, the most influential – the *Gentleman's Magazine*, estimated at 10,000.) By the end of the century, the print order for Moore's stood at 353,000 – a rise that continued to a peak of 560,000 in 1839.[31]

Vox Stellarum – or as it was widely known, *Moore's Almanack* – consisted of a simple ephemeris, listing the Moon's sign (by the part of the body it ruled), and the Sun's sign and degree, for each day of the year; lunar quarters, times of rising and setting, and major aspects to other planets; weather predictions, mixed with religious or political commentary, in the doggerel verse beginning each month or in a judicial astrological section; often a mysterious hieroglyphic; some simplified astronomy (concerning eclipses, for example); useful lists of tides, fairs, and so on; and a rough chronology of historical events. The aim was to provide guidance or enlightenment in all areas of major concern to its readers, throughout the year ahead: from the weather, agriculture and husbandry, to the larger events of politics, wars and natural disasters. This annual comprehensiveness and reliability, deeply comforting beyond the tabloid sensationalism of its astral revelations, marked out *Moore's* from the general body of popular prophetic literature in the eighteenth century.[32]

Its readership was clearly drawn mainly from the labouring classes. John Clare described the typical reading-matter in a rural cottage at the century's end as a Bible, a few chapbooks, *The Whole Duty of Man*, and an almanac. Clare vividly portrayed a typical farmer, perhaps from his beloved Northamptonshire, seated in the tavern and reading

Old Moore's annual prophecies
Of flooded fields and clouded skies;
Whose Almanac's thumb'd pages swarm
With frost and snow, and many a storm,
And wisdom, gossip'd from the stars,
Of politics and bloody wars.
He shakes his head, and still proceeds,
Nor doubts the truth of what he reads.[33]

In his autobiography, written (from a very different perspective) in the mid-nineteenth century, the publisher and reformer Charles Knight also recalled indignantly that

The believers in Moore's Almanack – and they comprised nearly all the rural population and very many of the dwellers in towns – would this year [1813] turn with deep anxiety to the wondrous hieroglyphic which was to exhibit the destiny of nations . . . There was scarcely a house in Southern England in which this two shillings worth of imposture was not to be found. There was scarcely a farmer who would cut his grass if the Almanack predicted rain. No cattle-doctor would give a drench to a cow unless he consulted the table in the Almanack showing what sign the Moon is in, and what part of the body it governs.[34]

Master Moore's readership was evidently not confined to the countryside. In the same period described by Knight, *The Stranger in Reading* (1810) recorded that after the Bible, the publication most read there was 'Moore's Almanack; this may be found not only in every house in the town, but also in every one in the neighbourhood, and partakes nearly of the same degree of belief in its prognostications as the Bible itself. Such,' the author concluded, 'is the ignorance of the lower classes of people here!'[35]

Often people wanted more specific and personal advice, on urgent matters, than was available from a book or almanac. Then they had recourse to the local 'wise' or 'cunning' man, or woman. While it is impossible to estimate numbers, it seems that this figure too had disappeared more from 'Books and Talk' than from 'the World'. Certainly in the more rural counties, or those relatively remote from metropolitan influence, he or she remained a recognized figure well into the nineteenth century, combining services of physic, divination and magical protection – all with a strong, if primitive, astrological component – for the poor.

One such was Timothy Crowther (1694–1760), a parish clerk in Skipton, Yorkshire, but better known there as an astrologer and magician. According to his nineteenth-century chronicler (who betrays the mixed scorn and fascination of the offended rationalist), 'People travelled long distances for the purpose of consulting him . . . He pretended [sic] to cast nativities, to predict events, to counteract the spells of the witch, to detect the thief and restore stolen goods, and to cast out evil spirits.' Crowther's manuscript of horoscopes and aphorisms was apparently an amalgam of his own work with that of Ptolemy and Cardano; it covered nativities, diseases and physic, elections (for when to undertake something), the significance of comets and eclipses, weather and meteors, agriculture and husbandry. His eldest son, Samuel, probably followed in his father's footsteps, and 'the phrase "as cunning as Crowther" was far into this [the nineteenth] century applied in Craven to one shrewder and cleverer than his fellows'.[36]

A somewhat similar character was John Worsdale, 'the Lincoln Wiseman or Astronomer', active from about 1780 till 1820; he must wait until the next chapter. But we should note, as yet another sign of the extraordinary endurance of popular astrology, the continuing presence of the cunning-man after 1800. 'Wise Man Wilkinson' lived in rural Yorkshire of the 1870s; his books consisted of 'old Latin works on the black art, or medicine, or mathematics [almost certainly astrology], or the distilling of herbs', and 'nothing that any clergyman or ministers could say could alter the popular belief in him'. Samuel Bamford, writing of his youth, recalled George Plant of Blackley, Lancashire as 'a firm believer . . . in the virtues of herbs under certain planetary influences; and in the occult mysteries of Culpeper and Sibly'. 'Oud Rollison' was another fortune-teller in rural Lancashire, 'with applicants for miles around', whose knowledge was based on works by Agrippa, Lilly, Gadbury, and Zadkiel. (It is tantalizing to learn that Rollison would sometimes meet with other 'wise men' in Manchester, choosing 'a quiet public-house' for the purpose.)[37]

Nor were they purely rural characters. As Robert Southey recorded in 1807, 'A Cunning-Man, or Cunning-Woman, as they are termed, is to be found near every town, and though laws are occasionally put in force against them, still it is a lawful trade.'[38] Despite the attrition among London's leading judicial astrologers, there always seems to have been (as one historian remarked) at least a 'small number of obscure professionals in London and elsewhere.' When William Byrd of Virginia visted London in 1718, he consulted several times a 'conjurer called Old Abram', probably an astrologer. A Mrs Williams could be consulted in Bath, Bristol and London in the 1780s, 'by ladies only'. William Gilbert, of Queen Square, London, specialized throughout the 1790s in the production of magico-astrological talismans, whose making he taught for a fee. Around the same time, there was a Mr Creighton, who 'used to be followed by great numbers, on account of his skill in astrology, and the medical art. He resolved questions for several years in Old Bailey, a few doors from Ludgate-Hill, on the right hand side of the way'.[39] In addition, there must have been others who kept their involvement more private – for example, Norris Purslow, a clothier in Wapping, London. Raised as a Quaker, and apprenticed at an early age, Purslow began studying Partridge and Ptolemy at the age of twenty-five. He kept an astrological diary covering the years 1673–1737, and apparently started a club of fellow aficionados in Tower Hill.[40]

This adaptation from countryside to town, which matched the general migration well into the nineteenth century, raises some difficult questions. If popular astrology constituted a 'mentality', essentially animistic and neo-Platonic, how did it survive the move to a relatively dispersed and utilitarian environment? If, on the other hand, it was merely a kind of technology, a rag-bag of specific remedies and responses, why was it not speedily replaced by a different and more 'appropriate' set? The answer is probably that it was both: the materially rooted mentality, or world-view, of labouring people with the desire to understand and modify (mitigate, facilitate) processes, both natural and social, whose effects, whether cyclical or sudden, could mean life or death, and were bound up with most experiences in between. Urbanization hardly ended this. And, as with every deep-seated outlook, certain specific courses of action were rendered plausible, even vital, at the expense of others. Thus, for its adherents, popular astrology offered a satisfyingly flexible mode of thought and action that was neither quite 'natural philosophy' nor 'religion' (nor even 'magic'), but something of each – just what most annoyed and baffled their respective authorities, with the time and inclination to pursue apostates.

On the practice of eighteenth-century popular astrology – and implicitly, the patrician view of it – I think the last word should be given to Richard Walton, a forgotten 'Student in Astrology and Physick,

Commonly call'd the Conjurer'. Walton was hanged at Warwick on 10 August 1733, for 'Promoting and Encouraging' two men to commit robbery. One he had reassured of favourable aspects on the night of the crime, apparently in ignorance of the thief's intentions; the other he had supplied with a small parchment of talismanic protection against witchcraft. In his 'Genuine Life, Confession, and Dying Speech', he recorded these plangent reflections:

> as for the Art of Astrology, some time ago I did daily practise it, in Calculating Nativities and setting Horeae Figures: Now whether the Art it self is an Imposture, or whether I have ever been of any Service to my Country by the Practice of that Science, that is not the case: . . . I may undoubtedly have been Serviceable to many People in a general way, and yet in particular Matters have deceived some and my self too. . . . I have some times made pretences of Calculating a Nativity for half a Crown, and that to[o], where the Age could not be given, no not to an Hour or two, when the Truth is, such a Work could not be done [for] under a Pound . . . Astrology is a Science no way fit or suitable for the Study of a poor Man, for except he be well accomplish'd, and hath a good collection of Antient and Modern Authors, he had better sit still.[41]

Attacks

The strength and continuity of eighteenth-century popular astrology is impressive, considering the hostility directed at astrologers by various authorities after 1660. Although direct attacks arguably subsided after the turn of the century, that antagonism did not. It had several strands. One was continued concern on the part of religious authorities. Charles Trimnell, Bishop of Norwich, inquired during a formal visitation in 1716 whether there was anyone in the parish 'that by Sorcery, Charms, or Astrology pretend to tell Fortunes or discover lost Goods, or any that consult with such Persons?' These persons were classed with those committing adultery, fornication, swearing, drunkenness, railing, and 'Sowers of Sedition, Faction or Discord among their Neighbours' – a revealing catalogue, which makes it very clear where popular astrology stood in the eyes of the Anglican Church.[42] The same attitude was confirmed in colourful terms by a later Bishop of Norwich (1790–2), George Horne. Referring to William Law's involvement with mysticism, and in particular the philosophy of Jacob Boehme, Horne wrote that he regretted Law's 'falling from the heaven of Christianity into the sink and complication of Paganism, Quakerism, and Socinianism, mixed up with chemistry [i.e. alchemy] and astrology by a possessed cobbler'.[43]

Another element was political. In 1708, an anonymous author anatomized the 'true character' of 'a Know-all Astrological Quack' (along with those of a 'A Female Hypocrite, or, Devil in Disguise. A Low-Churchman, or, Ecclesiastical Bifarius' and others). In his opinion, the astrologer should be placed among the Whigs 'that Fright the foolish World with Incredible Fears', and who, from 'Blazing-Stars [and] Dreadful Comets . . . raise such Predictions . . . as will gratify the Itch of a Seditious Party, and kindle Combustions among a Malcontented People, who delight to hear of Revolutions and Disturbances'.[44]

It is true that most of the contemporary almanac-writers were Whigs. By this time, of course, they were increasingly out of sympathy with the dominant and conservative Walpolian Whigs. The latter were anxious to distance themselves from any taint of insurrection or radicalism, and the questionable doctrinal aspects of the Glorious Revolution. By contrast, the almanac-writers themselves were often responsible for keeping alive an association with such memories; as late as the 1740s, the title-page of Partridge's old imprint *Merlinus Liberatus* described itself as appearing in 'the 54th [year] of our Deliverance by King William from Popery, and Arbitrary Government. But the 47th from the Horrid, Popish, High-Church Jacobite Plot'.

It also cannot have been coincidental that Jonathan Swift, the Tory dean of St Patrick's (albeit a disgruntled one), chose a leading radical Whig propagandist, as well as astrologer, for his satiric target. Significantly, however, Swift's lead was taken up by Joseph Addison – an important spokesman for the Whig party. In 1709, Addison and other literati, writing in the *Tatler*, warned readers to beware of people now claiming to be John Partridge. A few years later, however, a passage in the *Spectator* showed that their concern went well past a joke:

> It is not to be conceived how many Wizards, Gypsies and Cunning Men are dispersed though all the Counties and Market Towns of Great Britain, not to mention the Fortune-Tellers and Astrologers, who live very comfortably upon the curiosity of several well-disposed Persons in the Cities of London and Westminster. Notwithstanding these Follies are pretty well worn out of the Minds of the Wise and Learned in the present Age, Multitudes of weak and ignorant Persons are still slaves to them.[45]

In 1708, another critique of popular astrology appeared: the English translation of Pierre Bayle's *Lettre sur la Comète*. The significance of this book lies not in its effects – most likely a matter of preaching to the converted – so much as the nature of its case. Using the comet of 1680 as a springboard, Bayle used various rhetorical strategies in an attempt to discredit popular belief in comets as omens – all unashamedly elitist. One was the kind of wit and scepticism associated with the attitude of the literati, mocking the ignorance of the uneducated masses. But he also drew on recent scientific work (particularly in astronomy), carefully arranged to support an active but transcendental and non-magical God, whose proper interpreters were natural philosophers. In both cases, the object of scorn was 'superstition and idolatry', whether pagan or Catholic. This points to the guiding thread in Bayle's argument, which (far from constituting an undiluted rationalism, scientific or otherwise) was his commitment to Calvinism, always the most jealous among Christian sects of Gods's direct and unmediated will.[46]

Among popular scientific writers and lecturers, we find little direct mention of astrology. John Harris, writing in 1704, was content to dismiss it as 'a ridiculous piece of foolery'; Ephraim Chambers, in his influential *Cyclopaedia* (1728), as 'vain', and the later *Encyclopaedia Britannica* (1771) as 'a just subject of contempt and ridicule'. The continuity of outlook is clear (although in fact their treatment of astrology was more complicated than it appears here, as we shall see).[47]

Another writer and journalist, Daniel Defoe, used his *Journal of the Plague Year* (1722) to lambast the power of astrologers to spread panic, exploiting the superstitious credulity of the people – 'adicted to Prophecies, and Astrological Conjurations . . . Books frighted them terribly; such as Lilly's Almanack, Gadbury's Astrological Predictions, Poor Robin's Almanack and the like.' A few years later, Defoe published *A System of Magicke; or A History of the Black Art* (1727). The frontispiece showed a magician, conjuring in a circle of the zodiacal signs.

In 1740, the *Gentleman's Magazine* took up the earlier refrain of the *Tatler* and *Spectator* by lashing 'this absurd Chimera of judicial Astrology', and protesting 'the Impiety of the Practice, and the Improbability that the Decrees of Providence, and the Mystery of future Events, should be revealed to the most illiterate and profligate Fellows.'[48] In 1760, an Anglican divine added to Samuel Boyse's widely selling *Fabulous History of the Heathen Gods, Godesses, Heroes, &c.* 'by Way of Appendix, a rational Account of the various superstitious Observances of

Astrology . . . In which the Origin of each is pointed out, and the Whole interspersed with such moral Reflexions, as have a tendency to preserve the Minds of Youth from the Infection of superstitious Follies'.[49] Samuel Johnson, in his influential *Dictionary* (1775), tersely dismissed astrology as 'The practice of foretelling things by the knowledge of the stars; an art now generally exploded as irrational and false.' ('Superstition' he defined simply as 'False Religion.')[50]

A revealing episode took place in the 1770s. Thomas Carnan, a London publisher, won the legal right in 1775 to publish almanacs. Four years later, under Lord North, the government cynically attempted to pass a bill upholding the Company of Stationers' profitable monopoly; according to one observer, the Prime Minister 'wanted the stars, Moore and Partridge as backers in the American war.' This met with powerful opposition in the House of Commons, however. It was led by Erskine, a lawyer, who fiercely attacked the Company's almanacs. Significantly, he objected equally to the 'senseless absurdities' of Old Moore and the like and the coarse humour of the *Poor Robin*, from which Erskine offered to give some examples but desisted, since 'I know of no house but a brothel that could suffer the quotations'. The anti-astrological tenor of the latter was evidently no defence; rather, the entire almanac project was now objectionable, no longer as dangerous or seditious, but by virtue (so to speak) of its vulgarity. The bill was defeated, although the Company's monopoly was soon effectively restored by the imposition of a heavy stamp duty on its rivals.[51]

Thus, the link between astrology and popular profane magic, and between both of these and vulgar enthusiasm – against which the judicial astrologers of the late seventeenth century had fought so hard, and about which Coley had complained to Aubrey – was now firmly set, and astrology therefore 'exploded', among the classes of which these men were representative. Indeed, the consensus on the subject is striking. But their enlightenment was far from universal; we have observed the survival of popular astrology throughout the eighteenth century. This tenacity, and the gulf it implied between (broadly speaking) rulers and ruled, did not go altogether unnoticed by the former. I have already quoted *The Stranger in Reading* and the indignant Charles Knight. In 1803, George Beaumont, the minister of Ebenezer Chapel in Norwich, published an *Analyzation and Refutation of Astrology*. This was based on his shocked discovery that

A publication like this, I say, is really wanting; for though the present generation, like all preceding ones, has vanity enough to think, and frequently to call itself an 'enlightened age,' yet in spite of this conceit, and in spite of Societies for the Reformation of Manners – of Bible Societies – of Tract Societies – of Sunday Schools &c. there is in thousands and tens of thousands of people, a strong bias toward *Astrology and Fortune-telling.* This mischievous propensity is principally, though not wholly, among the lower ranks of Society.[52]

Even as late as the 1860s – a century and a half after the *Spectator*'s virtually identical complaint – folklorists found the situation apparently almost unchanged. In the words of two (which could be supported by those of others),

There is scarcely a town of any magnitude in Lancashire, or in one or two adjacent counties, which does not possess its local 'fortune-teller' or pretender to a knowledge of astrology, and to a power of predicting the future events of life . . . to a large and credulous number of applicants. The fortune-teller . . . professes to be able to 'cast nativities' and to 'rule the planets.' If, as in not unfrequently the case, he be a medical botanist, he gathers his herbs when the proper planet is 'in the ascendent.'[53]

There follows a description of a character instantly recognizable from any time in the preceding three centuries. This extraordinary resilience – along with the equally determined and persistent elite opprobrium – demands some further exploring and explaining.

From Popular to Plebeian

The general tenor of these attacks was consistent throughout the eighteenth century. In terms of learning, popular astrology was assailed as ignorance; in politics, as fanaticism; in religion, as superstition; and in general, as vulgar enthusiasm. Equally consistent was their social origin: the traditional religious elite, plus the new professional intellectuals (scientific and literary) – members of the non-landed professional and mercantile middle classes that appeared, in new numbers and influence, in post-Restoration metropolitan England. These groups, along with the traditional elite of the gentry and nobility, had by now developed a shared 'sense of separation of genteel from low mentality, and had fixed a broad range of associations – ignorance, superstition, sensuality, inferiority, violence, credulity, passivity – with the mind of the people.'[54] There can be no doubt that in the wake of the Restoration, popular astrology had taken up a secure place in this profane pantheon. One result was that it became an automatic (if often incidental) target for other groups, later in the eighteenth century, seeking to reform popular culture.

Religious criticism of astrology was hardly new; its greatest impetus lay in the Protestant Reformation (and the Catholic Counter-Reformation) of the late sixteenth century. The resulting assault on popular religion was manifested most clearly, in England, in the mid-seventeenth-century apotheosis of Puritanism; as we saw earlier, Presbyterianism (and Calvinism) were especially hostile to astrology. But it was equally plain that the restored Anglican Church (whether High or Latitudinarian) shared the same sentiments, if for somewhat different reasons. The concern voiced by the Bishops of Norwich simply carried them forward into the eighteenth century. That there were not more instances probably reflects the relative withdrawal of the church from the common people for several decades after the Restoration – and thus from their culture, where popular astrology lived on.[55]

Explicit attacks by natural philosophers in the eighteenth century were infrequent, and those I have quoted were contented to dismiss the subject in a few lines. The most plausible explanation is that by this time – and in so far as they too were less interested in proselytizing than in addressing an audience of peers and colleagues – natural philosophers regarded it as unnecessary to spell out the obvious concerning astrology. The distaste their few remarks and general silence convey is hardly surprising; this was an era when the membership of the Royal Society was nearly indistinguishable from that of the Antiquarian Society (with which it almost merged in 1729), and the leadership of both dominated by the Whig gentry.[56]

The most striking criticism of the early eighteenth century came from a new quarter: the leading metropolitan literati and 'wits', writing in the pages of the *Tatler*, *Spectator*, or *Gentleman's Magazine*. Like many of the natural philosophers their origins were overwhelmingly middle class, and their values – to a greater extent than those of any previous critics – secular and sceptical. But it would be highly superficial to accept that reading at face value. Consider the terms of Swift's condemnation, which he put in Partridge's mouth: 'I am a poor ignorant fellow, bred to a mean trade, yet I have sense enough to know that all pretences of foretelling by astrology are deceits, for the manifest reason that all the wise and learned, who alone can judge whether there be any truth in this science, do unanimously agree to laugh at and despise it, and that none but the ignorant vulgar give it any credit.'[57] The right to judge the truth is strictly reserved for the 'wise and learned alone', and denied to the 'ignorant vulgar'. Swift's construction, by which the former *agree* to laugh at astrology, is also peculiarly apt.

Indeed, the burgeoning periodical literature of the eighteenth century was more or less consciously constructed – as a model of how properly to think and write – in opposition to the popular almanac exemplified by *Moore*'s. The intended audience for the *Gentleman's Magazine*, and the nature of its authority, is plain enough from its title. By contrast, a correspondent fulminated in the *Derby Mercury* in 1817, 'lawless vagabonds . . . are so ignorant and infatuated, as to appeal for proof to the custom of the country, and the authority of their Almanacks.'[58] William Hazlitt summed it up thus: periodical literature, he wrote,

> does not treat of minerals or fossils, of the virtue of plants, or the influence of the planets; it does not meddle with forms of belief, or systems of philosophy, nor launch into the world of spiritual existences . . . It is the best and most natural course of study. It is in morals and manners what the experimental method is in natural philosophy, as opposed to the dogmatical method.

Compare how it looked to John Clare, writing at the same time but on the receiving end: 'grammar in learning is like Tyranny in government'.[59] This is what Terry Eagleton identified as the irony of Enlightenment criticism, 'that while its appeal to standards of universal reason signifies a resistance to absolutism, the critical gesture itself is typically conservative and corrective, revising and adjusting particular phenomena to its implacable model of discourse.'[60]

Eagleton identifies the literati as part of 'an historic alliance' between the English middle classes and their social superiors. The cultural project represented by the *Tatler* and the *Spectator*, while supposedly unpolitical, was nevertheless crucially sustained through political power; and if it 'was not especially political, it is in part because . . . what the political moment demanded was precisely "cultural".'[61] It may be objected that the literati were not of high social origin, and that, in addition to frequently disavowing politics, they could and did criticize aristocratic individuals or institutions. The Scriblerians, for example, mocked the wilder ideas of virtuosi in the Royal Society – but not its chief ornaments. Pope famously eulogized Newton, and both Addison and Johnson, for example, espoused scientific experiment and the new philosophy. Why? Because, as Locke had already said, these were fit subjects for a gentleman to study.[62] Similarly, patrician decadence, corruption and absolutism were also fair targets, but in a way which failed to raise questions about their fundamental legitimacy. As we have just seen, their attacks on astrology likewise reveal a far from independent, sceptical, or wholly secular outlook. Here, as in other respects, what emerges most clearly is the 'hegemonic voice of the gentry', whose Christian and loyalist aristocratic ethic was so unsettled by the enthusiastic (including astrological) excesses of the English Revolution.[63] In the wake of 1688–9, and into their long dominance under George I and II, that voice was at least equally likely to speak as Whig than as Tory; for both, radical Whig democrat–populists and almanac-writers, such as Partridge and Salmon, were beyond the pale.

A fresh assault, or rather assaults, on English popular culture commenced in the last third of the eighteenth century. One was religious, led by evangelical Anglicans and Methodists, filling the gap left by the Church's withdrawal from the people. Although Wesley eventually (and reluctantly) instituted his own ordinations, by the 1760s Wesleyan Methodism was becoming respectable, and beginning to close ranks with Anglicanism against the unreformed 'mob'. The breakaway Primitive Methodists, however, proselytized with aggressive vigour. To an unprecedented extent, this new dissent was both of and against popular culture.[64] Astrology (so far as I know) was rarely expressly targeted, but their fervent Christian monotheism left little room for rivals or variations; while the attack on popular recreations and festivals tried to undermine the implicitly pagan and popular-astrological sense of annual rhythms.

This reforming wave went hand in hand with another concerted attempt on the life of popular culture, this time secular.[65] Gathering pace towards the end of the century, and in tandem with fresh enclosures of commons (often the sites of fairs and festivals) to create profitable land-use, employers and manufacturers attempted to instil a new sense of labour-discipline in their workforce. Once again, popular astrology was not specifically under attack; a major part of its substratum, however, was: people's 'natural', cyclical and open-ended sense of time, including work-rhythms and holidays.[66]

Late in the century, still another group of reformers appeared: also secular and middle class, but politically and philosophically oriented. Bearers of the Enlightenment message, such as Jeremy Bentham and Joseph Priestley, took on both popular and patrician intransigence. Evidently they too did not feel that popular astrology as such was worth attacking. It is interesting to note, however, that when working-class radicals, such as Samuel Bamford and William Lovett, took up the cause in the nineteenth century, they invariably included an explicit repudiation of the popular astrology and magic with which they had grown up.[67]

Many reformers found common cause in a host of societies that sprang up at this time: the Proclamation Society (1787), the Religious Tract Society (1799) and the Society for the Suppression of Vice (1802) – 'the vices of persons whose income does not exceed £500 a year', as a contemporary caustically noted. Looking back to the Societies for the Reformation of Manners of the 1690s, and the Society for the Propagation of the Gospel of 1698, their members consisted largely of leading clergy, merchants and philanthropists, politicians and justices.[68]

Overall, these efforts were not particularly effective. They were not entirely without effect, however. For example, does astrology's decreasing mention by reformers mean there was less of it? The answer must be 'no'. The material I have already cited shows that popular astrology was widespread (if patchy) well into the mid nineteenth century. In addition, it was the target of a new wave of criticism early in that century – led this time by Charles Knight and others of the renewed Society for the Propagation of Christian Knowledge, the *Athenaeum*, and popular scientific writers. So we are confronted with repeated waves of concern, often in concert although differently motivated. Was English popular astrology, then, really unchanged from (say) the sixteenth century – a veritable mentality of *la longue durée*?

Again – although this question is more difficult to analyse – the answer is probably 'no'. Its apparent duration is uncontestable, a striking confirmation of the partial autonomy and 'authentic self-activity' of plebeian culture at this time.[69] (Indeed, the longevity of such beliefs may exceed what is generally known or admitted.)[70] Nor is that impression of stability wholly misleading. The figure of the cunning-man or -woman, the lunar lore, *Moore's Almanack*: these were extraordinarily durable. But there was a slow change taking place from the late seventeenth century to the end of the eighteenth. It is best described as a shift from popular astrology to plebeian – a process that began (allowing for earlier religious antecedents) in the late seventeenth century.[71] As we saw earlier, it was then that judicial astrologers began to feel the pinch, complaining, like Joseph Blagrave, that many former clients 'have refused to come or send unto me for help to cure their infirmities: and many of those who did come, came for the most part privately, fearing either loss of reputation or reproaches from their Neighbours, and other unsatisfied people.'[72] As the upper and then middle classes gradually withdrew from popular culture, and turned on it, its astrology became a purely popular, and therefore plebeian, phenomenon, rejected by polite and educated society, and thus isolated from other social and intellectual currents (outright hostility excepted). As John Rule has written of eighteenth-century popular culture as a whole, it was a 'survival' only in a qualified sense, because 'in the making of their own culture' – a making perforce more self-conscious, as a result of that desertion and the following attempt to re-mould them – 'the working people were defensive against repression and reactive against restraints . . . A defensive stance was inevitable.'[73]

This would explain the passively but determinedly resistant, static and repetitive quality of popular astrology later in the century – admittedly subjective, but perceptible – compared to the fructuous flux of a hundred years earlier. Perhaps it also partly explains the disappearance of figures such as Dee, Lilly, Partridge, and Salmon: astrologers whose skill and renown far exceeded the popular level, but who could comfortably move in and out of it, and speak for it. Contrast the surviving village astrologer, wholly and only in, and of, that culture. Even when the eminent judicial astrologers reappeared (generally speaking) in the 1790s – men such as Worsdale and Sibly – they held themselves aloof from their plebeian colleagues and clients, and addressed themselves to social and astrological elites.

Paradoxically, pressure from above – via 'increasing resistance and partial conformity from below'[74] – may have actually helped to maintain popular astrological elements within plebeian culture as a whole: for example, the success of *Moore*'s. It seems possible, at least, that his became the talismanic voice of popular astrology which could be relied on by labouring people, as they became increasingly a self-conscious, to unite and speak for them as a class. Certainly he was a model of consistency and continuity; the only variations throughout the century (and more) were familiar ones, appropriate to the year in hand. Without any contradiction, another factor was both the production and consumption of this almanac as a commercial success; that is, as a plebeian appropriation of Georgian entrepreneurialism, with the eager cooperation of the Company of Stationers.[75] It is, therefore, not too far-fetched to see in Old Moore the symbol of an enduring class mentality.[76]

Even here, however, the slow effects of confinement are apparent. For example, the original Francis Moore was, like his colleagues Salmon and Partridge, a radical Whig. To some extent, that remained true of his editorial successors in the eighteenth century; they opposed enclosures, welcomed American independence, and hailed the French Revolution as a herald of greater freedom.[77] On the other hand, they drew well back from overt republicanism in politics, and deism or atheism in religion. Underlying the often critical Whig points, there was a basic conservatism – professing (undoubtedly sincerely) loyalty to Christianity and the Crown – which reveals hegemonically determined limits. Other almanacs, too, showed an increasing inclination to avoid party politics; in the self-addressed words of one, 'Never concern thyself with Whig and Tory / Nor tell them of a court and country story.'[78]

Undoubtedly, this editorial stance reflected in large part a shrewd assessment, necessitated by the Company of Stationers' hard-nosed commercialism, of the labouring and lower middle classes who formed the almanacs' audience. Assuming that that assessment was correct, it may be seen as evidence that there was less plebeian support for contemporary radical agitation – as distinct from that of an activist middle-class minority – than is sometimes assumed. (The 'Church and King' mobs that burned down Joseph Priestley's house in 1791 were not particularly anomalous.) By the same token, popular astrology was evidently no longer a major part of oppositional discourse. None of the later eighteenth century radicals – Wilkes, Priestley, Godwin, Paine – used it as an interpretive or rhetorical resource. (Apart from the occasional more daring assertions in Moore's or Season's almanacs, virtually a lone exception was *The Celestial Telegraph, or Almanack of the People* (1795); written by one 'Astrologus', anonymity was probably well advised in view of the radical sentiments in the almanac and covering letter, addressed to the king.) All this was in striking contrast to the radical astrologers and their almanacs and books – consulted, employed and feared by all shades of opinion – in the mid-seventeenth century. The succeeding upper- and middle-class reactions failed to eradicate popular astrology; but they largely succeeded, by determined stigmatization and confinement to a plebeian ghetto, in depoliticizing it.

The 1790s in particular was a period of turbulence, marked not only by the appearance of English Jacobinism but also by an extraordinary plethora of popular religions. As J. F. C. Harrison has said: 'Political radicalism and religious millenarianism were not alternatives so much as different aspects of the same phenomenon.'[79] (The writings of Paine exemplify this point.) People were quick to relate the upheavals of the times to the large body of mystical and millenarian beliefs that was in one form or another common currency, impelled by that 'fascination with the odd, the mysterious, and the prophetic that permeated English culture in the era of the French Revolution.'[80] Among the instances of what one alarmed contemporary termed a 'Millenium of Infidelity', we find a resurgent Boehmenism, based on new translations of Boehme's works appearing in 1764–81; the spectacular mission of Richard Brothers, whose prophecies were then succeeded by those of Joanna Southcott; and a new occult religion, Swedenborgianism, with churches in London, Bristol and Salisbury.[81] Another such phenomenon was Mesmerism. Its English adherents were drawn mainly from the middle classes, including even some members of polite society and the literati; and Mesmer and his colleagues denied any connection between his supposedly physical 'subtle imponderable fluid' and astrology, occultism or mysticism. But they failed to convince the conservative scientific and medical establishment, and in the context of an anxious and Francophobic reaction in England to the French Revolution, the experts' poor opinions carried the day.[82]

One point that concerns us here is that these heterodox millenarian movements were not 'popular' in the sense of being pursued exclusively by the poor or illiterate. Rather, as Harrison points out, 'Such evidence as we have points to support from artisans, small farmers, shopkeepers, tradesmen, domestic servants and women, together with an important minority of merchants, businessmen, clergy and members of the professions'.[83] Such people also constituted just the constituency for popular astrology; and in the last half of the century, astrological almanacs did devote occasional issues to the discussion and comparison of biblical prophecies, current events, and astrological omens – often enough to indicate a real interest in the subject among their readers. Joanna Southcott presumably feared some competition from this source, since she warned her followers not to 'seek after stargazers or astrologers'. Not all her followers obeyed; a letter survives from one disciple to another, advising him to procure *Moore's Almanack*, and still another was moved to publish a testimony to the accuracy of its prophecies, in the 1750s and 1790s, concerning Southcott's ministry.[84]

Late-eighteenth century almanacs certainly discussed and disseminated millenarian ideas. While astrology was thus part of the ferment, it was, however, not – as it had been a century and a half earlier – a central part. The decades of suppression and exclusion had taken their toll. This point is confirmed textually: in their interpretations of prophetic texts from the Bible and overtly astrological interpretations of unusual celestial phenomena (comets, conjunctions of the greater planets, and so on), the popular almanacs were certainly more daring than the conservative, other-worldly Anglican millenarianism that followed the Restoration. But *Moore's* approach was also distinctly removed from the socially radical Puritan and sectarian millenarianism of the preceding century. Of course, it is also significant that Old Moore none the less lived on (and on). In short, as E. P. Thompson argued, the cultural hegemony of the gentry in eighteenth-century England, while not determining the content of plebeian culture, did exert influence, and set limits as to what was possible.[85]

5

Judicial Astrology: Decline

The Constituency

The practice of making relatively precise predictions or recommendations, based on the stars, was referred to in early modern England as 'judicial astrology'. Such advice or prediction pertained mainly to individuals. Its indispensable emblem was the horoscope, drawn up for a specific moment, at a specific place, which was then interpreted or 'judged'.

As a kind of middle-brow astrology, midway between popular and philosophical, it had two main distinguishing facets. On the one hand, it was considerably more complex, both astrologically (interpretively) and astronomically, than the former. The newest and/or most reliable ephemerides were usually used, and the positions of the planets were calculated with considerable precision, often involving mathematical procedures of no little difficulty. Practitioners usually needed to master such astrological niceties as 'azimuth', 'hyleg', and 'peregrine', in order to be able adequately to interpret the astronomy.[1] On the other hand, in common with popular astrology but as distinct from philosophical, the emphasis in judicial astrology was placed firmly on the lives of individuals, and their personal problems and decisions. (By considering the State or country as an individual, this could extend to wider political fortunes.)

These intellectual characteristics had social analogues. Judicial astrologers were distinguished from popular astrologers by their access to the degree of education (or self-education) required to undertake the drawing up and interpretation of a proper horoscope. As I have already mentioned, judicial astrologers were themselves aware of their removal from the general populace, including its simple, lunar astrology. While occasionally bemoaning popular ignorance, they were also often obliged to make concessions to it, such as simplifying their almanacs. Equally, however, they were rarely able to engage in full-time or professional natural philosophy, cosmology, and so on. Unlike those who did, judicial astrologers were not found in the Royal Society or the upper reaches of the Anglican Church, and their philosophical speculations – which remained identifiably astrological – were not considered worthy of serious consideration among those exalted ranks. (This point will become clearer when we proceed to the kind of astrology associated therewith, discussed in chapter 6.) Taken together, these characteristics serve to mark out judicial astrology and astrologers as relatively distinct types.

Did anyone in eighteenth century England still cast horoscopes? The short answer to this question is, undubitably, 'yes'. Overall, however, the eighteenth century was a low point for judicial astrology – dramatically, when compared to the preceding century, and noticeably also compared to the following. Decline is evident, for example, in the number of publications, and (where ascertainable) the number of copies sold. Indeed, the first such book in the century by an English astrologer, David Irish's *Animadversio Astrologica* (1701), sounded a defensive note:

> Thus you see, [even] if it were granted, that the Stars are the Causes of all sublunary Alterations, how difficult, if not impossible, it is to guess aright at future events by them . . . In my Opinion, if anything Astrological be useful beside Physick, 'tis by the Observation of the great Eclipses of the Luminaries, Conjunctions, Commets, and Blazing Stars, which, though with great uncertainty to Time, forewarn Mortals of future Destruction, Revolutions in Kingdoms or Calamities impending.[2]

This marked a retreat from strictly judicial astrology (except in its medical applications) towards the vaguer and more generalized 'natural' astrology of the common people, and, in a very different context, the natural astrology of philosophers such as Whiston. The same kind of retreat was perceptible in John Gadbury's final almanac of 1703, although he had disparaged 'pompous predictions' for many years. Later in the eighteenth century, Henry Season too derided precise predictions as astrologically heretical and 'crude'.[3]

Another book discussed the astrology of a solar eclipse in 1715, noting that 'Dr Flamstead, Mr. Halley, Mr. Wiston, and several other astrologers [sic], differ very much in their conjectures of what may happen to this Kingdom.' It concluded, cautiously and comprehensively:

119

'God preserve England from Plague, Pestilence, Famine, Battle, Murder and Sudden Death.'[4] Books of astrological prophecy based on horoscopes also appeared in 1722 and 1733: two in the former year, concerning a conjunction of Mars, Jupiter and Saturn, and a solar eclipse.[5] These show that there was still a literate market for specifically astrological prophecy; but there had been a real decline compared to the number of such publications almost a century earlier.

When we consider astrological textbooks, a similiar picture emerges. There is no way of knowing the number of copies sold, but the number of books actually teaching judicial astrology that were published between 1700 and 1790 – and excluding one generalized defence, in 1785 – was only six. In addition, that figure included two new editions of seventeenth-century textbooks, and a new edition and translation of Ptolemy's *Tetrabiblos*. This was markedly down from the mid- to late seventeenth-century peak, especially in the light of the dramatic rise of English publications generally after 1700. It should be noted, however, that there was always at least one textbook available.[6]

Among almanacs and ephemerides, the success of *Moore's Vox Stellarum*, which overlaps somewhat with judicial astrology, has already been described. But no one could have learned or practised judicial astrology from *Moore's* alone. It is interesting, therefore, to notice the survival of more technically complete almanacs and ephemerides, giving information on the planets' positions that was sufficiently precise to enable a horoscope to be calculated. Furthermore, such publications must have been profitable; they were printed and sold by the Stationers' Company on a strictly commercial basis.

One such almanac was Parker's *Ephemeris*; appearing annually from 1690 until 1781, it gave both geocentric and heliocentric positions of all the known planets for every day of the year. In 1773 (for example), 2,000 copies were published.[7] Other examples of this genre are Edmund Weaver's *The British Telescope* (1723–49), and Robert White's *Coelestial Atlas*. The latter began in 1750 and continued, under a succession of editors, into the nineteenth century. It is especially notable both for its wealth of astronomical detail, and for the complete lack of sensationalistic predictions that might otherwise explain its purchase. (White himself forswore all 'fallacious Prognostication'.) The *Atlas* also published 2,000 copies in 1773, and after absorbing the circulation of Parker's ephemeris in 1781, it sold 4,500–5,000 copies a year for the rest of the century.[8] It is difficult to believe that these entire sales were absorbed by amateur astronomers – especially in the light of other evidence pointing to judicial astrology. At the very least, we should again note that the necessary tools of the trade were available throughout the century.

In the same period, however, a new phenomenon appeared: highly literate and numerate but decidedly non-astrological almanacs, containing a wide variety of information on the secular and Church calendar for the year, plus a miscellany of other items, including mathematical games and puzzles. This trend began in 1684, when Richard Saunder assumed control of the previously astrological *Apollo Anglicanus*; he continued to produce it until his death, and included such essays as 'A Discourse on the Invalidity of Astrology.' But the most striking success was the *Ladies Diary*, started in 1704 by Coventry schoolmaster John Tipper. Addressed to educated middle-class women with time to spare – but probably selling to a wider readership – it reached a peak of 30,000 copies in 1750 (from which it gradually declined). (A related imitation dating from 1741, the *Gentlemans Diary*, was less successful.) These figures are impressive, though never approaching those for *Moore's*. Interestingly, even here the Moon's phases and the ingresses of the Sun into each sign of the zodiac were still given.[9] Overall, then, we notice with almanacs the same trend as with astrological books: attrition, without ever reaching the point of extinction.

Judicial Astrologers

Sometimes, as with Saunders and White, almanacs and ephemerides were edited by critics of astrology. But such men were the exception rather than the rule. So who were the astrologers who compiled and wrote the 'better' sort of almanac? In general, they were of middling social origins. Educated men, often autodidacts, they were well-read in the scientific literature of the day, and equally active as surveyors, mathematicians and amateur astronomers. To a striking extent, they lived in a well-defined area: the East Midlands (including what used to be Rutland), Leicester-shire and South Lincolnshire. It is worth recalling that both William Lilly and Richard Edlin, prophets of the plague and Great Fire, hailed from this area.[10] We have also encountered their contemporary, Vincent Wing (1619–68), the respected astronomer-astrologer of Luffenham, near Stamford. He sired a virtual astrological dynasty in Rutland, stretching well into the eighteenth century. Its members included his son Vincent II (1656–78); his nephew John (1643–1726) and John's son Tycho (1696–1750); and Tycho's sons, John II (1723–80) and Vincent III (1727–76).[11]

Tycho Wing was a surveyor and teacher of mathematics, living in Stamford and Pickworth. During the 1730s and 1740s he edited his own, Coley's, Andrews', and Moore's almanacs, maintaining a high standard of both astronomy and astrology. Vincent II and III also edited or helped edit almanacs. Vincent III worked with Thomas Wright (c.1716–d.1797), a surveyor and mathematician of Eaton, near Melton Mowbray, Leicestershire. In the 1760s and 1770s, Wright edited Moore's, Partridge's, Wing's and Andrews' almanacs. Edward Sharpe (fl.1730–73) was a physician in Stamford, who helped edit and advertised his astrological physic in several almanacs in the 1760s. William Harvey (fl.1750) of Knipton, Leicestershire, probably edited Wing's almanac after the death of Tycho. Edmund Weaver (fl.1723–d.1748) was a physician and surveyor of Frieston, near Boston, who compiled *The British Telescope* (in which he also advertised his astrological services) between 1723 and 1749.[12]

These astrologers carried on seventeenth-century traditions of scientific popularization by astrologers – covering topics such as the periodicity of comets, gravitation, meteorology, and the discovery (in 1781) of Uranus – and the combination of judicial astrology with astronomy, mathematics and surveying.[13] The astrological element was clearly sincere, centring on a belief in at least the possibility of a valid judicial astrology. Concerning their science, these astrologers were unquestionably 'enlightened' and informed; but unlike their predecessors of a century earlier, they no longer included members at the cutting edge of contemporary scientific work. Their mathematics were also different: still of a high standard, but more applied and observational, and less theoretical, than that of their forebears.[14]

This community (or more accurately, network) of judicial astrologers was not restricted to compilers of almanacs. One member was the eminent antiquarian, William Stukeley (1687–1765), MD and FRS. Born in Holbach, Lincolnshire, Stukeley was the son of a country gentleman with a small estate. He studied medicine in Cambridge and London, but gave it up in 1729 to become a divine in the Church of England. He divided the rest of his life between his native Lincolnshire and London, with benefices in both (finally as Rector of St Georges in Queens Square,

London). Stukely was passionately interested in ancient history, archaeology, and natural philosophy. He took up and developed John Aubrey's earlier interest in the stone monuments at Avebury, and engaged in cosmological speculations with Whiston, Bentley and Derham. Like others at the time, he was fascinated by the question of the locations of Heaven and Hell; where Whiston held that sinners will one day be relocated to a comet, however, Stukeley interpreted the band of the Milky Way to show that the Sun and stars lay on a plane dividing Heaven (above) from Hell (below).[15]

These pursuits included judicial astrology. Stukeley was a good friend of both Tycho Wing, 'with whom I spent many agreeable hours', and Edmund Weaver – 'we often agreeably entertained ourselves in calculations of astronomy, with a view to antient history'.[16] In fact, his interest can be dated to 1735 (aged forty-eight), when Tycho sent Stukeley his horoscope. 'Tho' I am a stranger to Astrology, and therefore cannot pretend to say anything for or against it: yet in compliance to my friend Wyng I considered his table of Directions and compared them with the memoir of my past life.' Stukeley was impressed: a long square to Saturn had corresponded to many difficulties, including a malicious lawsuit and being passed over for three clerical preferments. Stukeley also noticed that the same aspect had applied 'when I was in utmost danger of being ruin'd by the South Sea' (the financial crash). 'Upon these misfortunes sufficiently verifying my friend Wyngs predictions, I began to have an opinion of his art.'

Stukeley became deeply interested, studying his own horoscope and compiling a list of axioms – accompanied by a complex diagram – for an 'EIMAPMENH, or Rational Astrology'. Like John Aubrey and Robert Burton earlier, he viewed judicial astrology as essentially the science of good or bad Fortune: 'and if as Christians ought to doe, we call it the direction of providence, I see no difference at bottom.' In addition to prophecy and ordinary wisdom, Stukeley felt it reasonable to surmise that God had provided us with 'a kind of book of the whole series of general events that are to happen to each man . . . as far as it is consistent with mans free will of action.' As evidence that there 'may be somewhat in the art, more than what the generality of the learned now allow', Stukeley cited the phenomena of menstruation and gestation, 'the attraction of the heavenly bodys, electricity, magnetism, the flux of the tides, Dr Mead's book De imperio solis & lunae, and very many appearances in nature', together with – and this is a long step in the direction of judicial astrology

– 'many of the astrological aphorisms which have been verified by long experience'. Echoing Gadbury, he maintained that

> These matters give hints not obscure, that probably those learned men who decry alltogether any notion of Astrology, without understanding it, are to blame: as much as those who are too fancifull and credulous in admitting all the dregs of it, voyded by ignorant sciolists.

As his list makes clear, Stukeley was anxious to base judicial astrology on rational and natural grounds. Like earlier generations of astrologers in the same predicament, this led him to reject astrology's 'vulgar and seemingly irrational jargon', and concentrate on the observable and measurable planetary aspects. Having assured himself of judicial astrology's religious and scientific probity (in principle, at least) Stukely felt free to do so. Indeed, he argued (like Boyle) that 'it becomes us to examin into it. Otherwise astronomy and its modern improvement and scrupulous nicetys in observation, seems no more than speculation and curiosity, with very little use'.

Stukeley was in the problematic terrain familiar to us from the late seventeenth century, when it became imperative for judicial astrologers to differentiate themselves from the profane (whether religiously or philosophically) magic of popular astrology – but exceedingly difficult to do so, given their attachment to a judicial dimension. Unlike natural astrology, especially when armoured with Scripture and experiment, judicial specificities were highly vulnerable to being construed as divinatory and therefore suspect (enthusiastic, vulgar). Hence the anxious efforts to maintain otherwise. Despite such problems, however, this 'middling' astrology – wavering between the illicit or openly magical and the licit, theologically sanctioned natural philosophy – held a profound appeal for men such as Stukeley: 'There is something in it so agreeable to nature, to the chequerwork of life, that . . . it strikes a considering mind with great pleasure: and tempts us at least to wish it were agreeable to truth and fact.'[17]

Later in the eighteenth century, another member of the same network became its leading, if modest, exemplar. Henry Andrews (1744–1820) was born in Frieston, near Grantham, of poor parents. After peregrinations around Lincolnshire, during which his education was paid for by various employers, Andrews moved to Royston, Hertfordshire, in 1766. The following year, he opened a school (eventually a boarding school).

Word of his skill in mathematics and astronomy began to spread, perhaps through *The Royal Almanack*, he started in 1776, and he was employed by the Board of Longitude to compile the *Nautical Ephemeris*. This forty-year service was paid only nominally, but he received 'a handsome present' upon retiring. Andrews corresponded at length with the Astronomer Royal, Nevill Maskelyne, who evidently recognized the former's abilities. Upon his death, Andrews even received a respectful obituary in that occasional scourge of astrology, the *Gentleman's Magazine*.

As its author allowed, Andrews had also written and compiled *Moore's Vox Stellarum* for over forty years (roughly from 1783 until his death). In so doing, he had undertaken 'this task for the Company [of Stationers] and conformed himself to the rules of astrological art, without laying any stress upon his predictions, or taking any credit for the chances which sometimes led to their fulfilment.' Nevertheless, predictions Andrews certainly made, and freely chose to make. (The reason cannot have been renumeration; with its customary parsimony, the Company paid him a flat fee of £25 per annum.) They were freely mixed with political and scriptural comment, practical information, popularized science and usually at least one mysterious hieroglyph. As Capp remarked, the central section of *Moore's*, entitled 'Astrological Judgement', was 'in effect a synthesis of the author's personal convictions, biblical prophecy, and astrological predictions. There is no reason to doubt its sincerity.' Nor Andrews's success; he injected new life into *Moore's* pages, and, during his sway as editor, sales increased from 100,000 to 500,000.

Andrews's political and social outlook was that of a progressive Whig, who welcomed the American and French Revolutions and worried about royal and religious tyranny. On the other hand, he defended the Protestant monarchy, and such Anglican lynchpins as belief in the Trinity. (He was a Churchwarden in Royston for three years.) Despite a wide readership, his views were probably less influential than those of his predecessors; gone were the days when Lilly's predictions were debated by Parliamentary committees, when Gadbury was burnt in effigy in the streets of London, and Partridge obliged to leave the country. In relative terms, the audience of the judicial astrologers, even the most widely read, was now both smaller and more restricted. Nonetheless, Andrews spanned a remarkable social and intellectual spectrum. He was read and debated by thousands every year, with no less respect than he received from the Astronomer Royal. Perhaps more typical, however, was the young woman engaged to wait upon him in his last years, who many years later reverentially recalled: 'Every thing that he said would come to pass, has come to pass.'[18]

Judicial astrology in the eighteenth century was not confined entirely to sons of Leicestershire and Lincolnshire. Indeed, perhaps a better choice for a 'radical' astrologer would be Henry Season (1693–1775). This Wiltshireman was a licensed physician who published his own almanac (with annual sales of about 3,000) from 1733 until his death. His astrological and political convictions were equally strong, and he defended both in articulate and uncompromising terms. Despite hard words about the ignorance of the rural poor (whom he also variously compared to Hottentots and monkeys), he also defended them, attacking the government's callous complacency about their hardships and poor health. A friend of the radical Whig the Earl of Shelburne, Season praised Wilkes and Pitt, and ceaselessly criticized the corruption of peers and ministers.

Season's views on astrology were also interesting. He eschewed arcane considerations, such as the significance of individual degrees, planetary hours, and 'smoaky', 'lame', and 'azimine' placements – ''tis meerly Arabian Foolery, having no Basis in Reason or Experience for it'. But despite the similarity of his rhetoric to that of Partridge, he also attacked Placidus's 'rattle-brain'd Whim', adding that 'some popular Errors . . . have been confidently supported by adhering too much to the Doctrines planted by Ptolemy, taking all he wrote for Infallibility; for as he was deficient in his Astronomy, so consequently he must be the same in his Astrology too.' Since he also cited Boyle's concept of the planets' influence on the atmosphere, and hence on people, it seems that Season's rationality was in a scientific vein. However, despite his ambitions 'to restore and recover that antient, and excellent Art of Astrology, to it former Purity and Strength' – presumably with the help of natural philosophy – Season fell far short of doing so. His predecessors Gadbury and Goad, half a century earlier, had come closer, when the gap was not yet so wide or deep.

Even here, in the words of an avowed astrologer and radical, we find the deep impress of Anglican and scientific propriety. Nonetheless, Season's own integrity also shines through:

> Men abuse the Stars, who make use of them in any magic Spells, and all fortilegious Divinations whatever, where no Reason can be produced from

the Cause to shew the Effect. The Stars cannot effect any Thing on Man's Will directly, but *ex accidente* . . . nor doth any Christian Astrologer maintain or support such Practices. But by the Ignorance of some, and Superstition of others, these vile Uses have crept into Astrology; and 'tis from this the Enemies of this Art have condemned it, and vilified all that profess Astrology; and many others, merely from blind Tradition, do ridicule an Art they are entirely ignorant of. The ancient Astrology is a Branch of natural Philosophy, no way inconsistent or impugned by Scripture . . . All Eclipses, great Conjunctions, Comets and Configurations of Planets, are all Oracles of Divine Providence; which, whosoever despiseth, contemneth the Admonitions of God.[19]

The undying (it seems) hope of the reforming judicial astrologers now fell on entirely deaf ears. But the last note is one with which — especially respecting comets — many people of his time, quite unwilling to 'profess Astrology', agreed with Season. (That kind of astrology, if such it be, is the subject of the next chapter.)

Middling Astrology and the Middle Classes

As we have seen, committed judicial astrologers in the eighteenth century tended to be men of some, even considerable, learning, but no great social rank — in short, professional people of the 'middling' sort, and therefore at some (if no great) remove from the contemporary epicentres of power, influence and prestige. The picture of those not themselves astrologers or almanac-writers, but who none the less availed themselves of judicial astrology, is naturally more shadowy. We may at least conjecture that it was composed of people of similiar social and intellectual standing, whose numbers were considerably fewer than those readers who relied solely on Old Moore.

In addition to what we know from William Stukeley's interest, this picture is sharpened somewhat by other surviving diaries. For example, on 12 October 1717, the historian and librarian Thomas Hearne (1678–1735) (a sharp critic of Stukeley, incidentally) recorded that he had been visited by a Mr. Hayward of Garford, with whom he discussed ancient coins. 'This Hayward,' Hearne noted acerbically, 'is a mighty Admirer of Astrology, and particularly of the works of Wm Lilly, the Figure-Flinger.'[20] In Shropshire, Richard Gough (1635–1723), an educated yeoman, related the deaths of both James II and William III (in 1701) to

127

a solar eclipse the preceding year. He based this on his own interpretation of the eclipse, according to Greek and Arabic astrological rules. He comforted himself, in conclusion: 'Deus Astra regit.'[21] William Dyer (b.1730) was an accountant and philomath, living in Bristol. According to Jonathan Barry, he thought 'that the regular abatement of his father's illness after sunset was clear proof of the influence of the planets, including the sun, on the body'.[22] James Woodforde (1740–1803), a Norfolk parson, recorded his purchase of Moore's almanac in 1781. He related his health generally, and epilepsy in particular, to the phases of the Moon – recording for 29 May 1801, for example: 'I was very poorly myself all Day. I daresay the full Moon much affected us.'[23] Finally, Samuel Johnson's confidante Mrs Hester Thrale (1741–1821), was of poor gentry background. In a manner similiar to that of Henry Andrews, she freely mixed civil events, cosmic phenomena, and Biblical revelation. In 1792, for example, she related 'the King of France's brutal murder' to both 'the Thirteenth Chapter of St John's Apocalypse', and 'A Spot in the Sun large as Venus in her Transit.'[24]

These instances confirm the impression that in eighteenth-century England, judicial astrology was now confined to a small minority among the educated and professional middle classes. It is possible, however, to be more specific. Nearly all the individuals just mentioned also have in common a social and/or geographical distance – often by virtue of their provincial location – from London and its social and intellectual dominance. (In Mrs Thrale's case, her anomalous position as a woman intellectual and writer, of which she was well aware, probably introduced an analagous degree of social marginality.) This finding is readily understandable: it was first and foremost in London that the rise of the middle classes took place, marked by a close identification with the attitudes and values of their social superiors.[25] Given the elite view of judicial astrology after the Restoration, its social and geographical distribution now falls into place. In the course of the eighteenth century (as Barry has recently suggested),[26] significant differences seem to have developed between the metropolitan and provincial middle classes, with the influence of upper-class ideals and ideology penetrating the former more deeply than the latter. (A provincial impetus to imitate and absorb metropolitan culture only became irresistible in the nineteenth century.)[27]

This theory would clarify several things: for example, why there was such a paucity of judicial astrology in eighteenth century London, where the literati so comprehensively rejected it – unlike its cautious acceptance by a rural parson such as Woodforde, or provincial town-dweller such as Dyer, despite sharing the same basic class membership. It would also explain why the sharp rise in the numbers of those of middling status, wealth and profession was not reflected in the constituency of judicial astrology; this rise was primarily a metropolitan phenomenon. Above all, the flowering of judicial astrology in Leicestershire, Lincolnshire and the East Midlands (if not its apparent absence in other similarly placed counties elsewhere) becomes explicable. Lincolnshire in particular had a large population living in open parishes, interspersed with market towns, and on both accounts relatively independent of a largely weak or absent gentry. Attempts to drain the fens in the late seventeenth and eighteenth centuries encountered stiff, sometimes violent, resistance, which was led by 'the "middling sort" of the fenland villages – minor gentry, yeomen, richer husbandmen, some tradesmen'.[28] It cannot be wholly coincidence that this lack of deference characterized the same class, and in the same place, that sustained judicial astrology throughout the difficult decades of the eighteenth century.

In the nineteenth century, as noted above, the provinces strove to assimilate metropolitan values, perceived as sophisticated and exciting. This was not without earlier precedents. One marker, very late in the eighteenth century, was the widespread appearance of 'Lit. and Phil.' societies in the larger cities. These coincided with the relative integration of natural philosophy into the dominant and elite culture that was now beginning to colonize the provinces. In the process of its absorption by the middle classes, a lucrative lecture-tour circuit for scientific popularizers came into being. And there was a boom in the publishing of popular natural philosophy: versions of Newton's works, textbooks, encyclopaedias, and scientific literature for women (for example, the *Ladies Diary*) and even children.[29] These publications reached an increasingly wide audience, especially in the many coffee-houses and 'penny universities'. Boyle lecturers, such as William Derham and John Harris, managed to combine the two, speaking and publishing, to good advantage. The impact on judicial astrology, in so far as its natural constituency was (potentially) among the same readers, must have been considerable, without necessarily any direct mention of the subject at all.

In such works that did mention astrology, the approach was rational and experimental, while the tone seems increasingly secular, even implicitly deist, under the aegis of Anglican respectability. Such an approach effectively undermined two principal ways of conceiving and justifying astrological effects: as magical operations of sympathy and antipathy, and as manifestations of neo-Aristotelian rationalism. A third way adopted by astrological partisans was to view such influences as explainable (and therefore justifiable) in terms of experimental natural philosophy. Contemporary 'popular' scientific literature approached that possibility in a complex and cautious way. Sometimes, there was simply a blanket condemnation. More often, however, astrological ideas that seemed useful were accepted, but carefully qualified and whenever possible re-described in natural philosophical terms.[30] In this strategy, and the divided responses overall, it is possible to detect a residual ambivalence about astrology (and one that is missing, in the same kind of literature, a century later).

As in so many other ways, such as the popular millenarian movements we reviewed earlier, the 1790s were marked by upheavals in judicial astrology. Serious cracks appeared in the social and intellectual domination of the gentry, which had hitherto so successfully kept it down. One straw in the wind was the appearance of the first periodical ever to be addressed primarily and explicitly to other astrologers: *The Conjurer's Magazine*, later renamed *The Astrologer's Magazine*, which appeared from 1791 till 1794. This extravagant *mélange* of horoscopic interpretation, occult philosophy, physiognomy, mesmerism and Nostradamus constituted a package that is immediately familiar to the modern reader from certain widely selling magazines today. Such a novel attempt – itself short-lived, but a forerunner of others soon more successful – was a sign of the successful struggle, beginning in the 1790s, for greater intellectual independence by members of the middle classes. Significantly, this was not another plebeian or provincial almanac; it was a metropolitan astrological publication, by and for relatively educated readers.

Another point of interest is the way in which *The Conjurer's Magazine* articulated its independence. One of the contemporary touchstones of orthodoxy – despite its own initially controversial theological implications – was Newtonianism. Newton himself, as eulogized by Alexander Pope, had become a national hero well before his death in 1727, and his state funeral marked an apotheosis, but not an end, in the process of canonization. As the *Encyclopaedia Britannica* declared in 1771, in legitimate natural philosophy the place of honour was held by 'the only true and rational scheme, restored by Copernicus, and demonstrated by

Sir Isaac Newton.'[31] (Thus, for example, the leading English disciple of Jacob Boehme at this time, William Law, tried to sanctify Boehme by showing that he had anticipated and influenced Newton.)[32]

The ideological importance of Newtonianism in the eighteenth century is also perceptible through the opposition it aroused. It was hated and reviled, for example, by High-Churchmen and ultra-Tories whom the long Whig dominance had displaced; these men rallied around the conservatively Christian natural philosophy of John Hutchinson and Roger North.[33] At the other end of the spectrum – among the sectarians and millenarians who flourished in the 1790s – Newtonianism provoked similarly strong feelings. Mesmerists, intellectually and socially ambititious, and hopeful of advancement, wanted not to replace his theory but to supplement it with additional physical laws pertaining to their proposed 'subtle fluid'. By contrast, most of the popular millenarians – who had little hope of finding a place within the establishment, and desired rather to supplant it themselves – simply attacked such laws wholesale. Richard Brothers, for example, upheld the literal truth of Genesis, and assailed heliocentrism as 'erroneous, wild, and unnatural'. One of Brothers's later correspondents was a Liverpool astrologer, Bartholomew Prescot, who wrote in defence of a geocentric cosmos.[34] And in 1792, *The Conjurer's Magazine* informed the ghost of Newton that he 'may keep his nonsense of vacuum and attraction out of the way, for we are not indebted to mythology for life and presiding genii in the sun, planets, and all creation, but to sound reason, genuine theosophy, and the oracles of GOD'.[35]

It appears therefore that by this time Newtonianism was thoroughly embedded in contemporary dominant ideology. Perhaps as a consequence, there was also a marked difference between these judicial astrologers' antagonism to established natural philosophy, and those of the late seventeenth century, who had been in the forefront of attempts to advance and popularize it (and heliocentrism in particular). In fact, that tradition was maintained throughout the eighteenth century by the judicial astrologers of Rutland, Leicestershire and Lincolnshire. The key difference lies in the necessity that their metropolitan counterparts now felt (and the opportunity they saw) to break away from a hegemonic grip which, further from London, had never been as strong in the first place.

Worsdale and Sibly: Judicial Astrology 'Reborn'

Late in the eighteenth century, two judicial astrologers appeared who transcended their immediate locale through writings which approached the national circulation (if not the influence) of Lilly, Gadbury and Partridge. One was a product of the astrological milieu in the north of England that we have discussed; the other, unusually by this time, came from the south. Their sharp differences are belied by the conditions that permitted the relative success of both, namely the seismic contemporary changes in English society which opened up such new possibilities.

John Worsdale (1766–c.1828) was born in Fulbeck, near Grantham. After working there as a parish clerk, he lived in Helpringham, Spanby and Donington before settling in Lincoln, 'near the Cathedral'.[36] He was the upholder of a venerable astrological tradition, one which, in his opinion, Sibly betrayed: that of judicial astrology as a science neither magical nor modern, neither occult nor mechanistic, but based on rational Aristotelian principles and mathematical procedures. His predecessors in this, whose lead he acknowledged, were Partridge, Whalley and Placidus. Like them, he viewed the science of Galileo and Newton as, at best, irrelevant, and, at worst, a destructive aberration; accordingly, he viewed the past efforts of Gadbury and others to place astrology on a more scientific footing as merely clumsy meddling with tradition. But he was equally opposed to a magical astrology, which offended his view of astrology as a rational discipline, employing the formal canons laid down by Ptolemy and refined by Placidus. It also was offensively populist and heterodox. (Despite his own unorthodoxy – God had His place, but only as First Cause – Worsdale opposed 'Infidels, Deists, and Atheists'.) Worsdale, accordingly, was an unrepentent elitist. The dedication of one of his books (to Henry Andrews) remarked: 'I do not require the Vulgar and Illiterate to busy themselves with a subject of this nature; it is to you, Sir, and those that are learned in the Sydereal Mysteries, I only appeal in this case.' And he emphasized that a great deal of study is necessary in order to become 'qualified to decide, whether a planetary influence operates among mankind, or not'.[37]

In keeping with this tendency, Worsdale's judicial astrology involved highly detailed mathematical calculations, particularly concerning directions, and arcane interpretive points without any possible physical rationale, such as the apheta and hyleg. In several of his books, he

132

remorselessly recorded the putative accuracy and reliability of these techniques, the subject meeting his or her predicted death with ineluctable punctuality. This was his favourite test, and comprised the entirety of, for example, his *A Collection of Remarkable Nativities . . . Proving the Truth and Verity of Astrology, in its Genethliacal Part* (1799). Perhaps it was his revenge on the clients who supplied him with most of his livelihood, but for whom he clearly had scant respect. Like Partridge, Worsdale rejected horary astrology on principle, as rationally indefensible; but equally, it is unlikely he scrupled over its use in practice. Worsdale was not the only astrologer in the area; he himself complained about 'illiterate pretenders, who too often deceive the ignorant and unwary', adding that 'There are many of this tribe in several places, and there are some also that reside in the City of Lincoln, who, in alehouses and other noted places, prate about calculating Nativities.'[38] He certainly seems to have been the best-known astrologer, however, and John Worsdale, or Worsdall, 'the Lincoln Wiseman or Astronomer', survived well into the nineteenth century as a figure of popular folklore. In this capacity, he had uncanny powers, detecting thieves and locating lost objects for farmers, gamekeepers and servants, who regarded him with uneasy respect. He was even invariably accompanied by a familiar spirit, usually a blackbird or black cat. Thus did the common folk he despised obtain their unintentional posthumous revenge.[39]

In addition to his *Collection*, Worsdale published *Genethliacal Astrology* (1796), an analysis of *The Nativity of Napoleon Bonaparte* (1805), and (echoing Placidus) *Astronomy and Elementary Philosophy* (1819). In the last, he ventured the opinion that the 'elevated and dignified' positions of the planets at the time of American independence 'most clearly forebode, that the time will arrive, when THAT EMPIRE shall give laws to all Nations, and establish FREEDOM and LIBERTY in every part of the habitable Globe'.[40] His last book, published after his death by his son, was *Celestial Philosophy, or Genethliacal Astronomy* [*c*.1828]. There Worsdale continued to boast of his accuracy in predicting his critics' and clients' deaths – often to their face. He also reaffirmed his unwavering faith in Ptolemy, and lashed past and present astrologers who fell short of his particular and demanding standards: Gadbury, Coley, Parker, and the new generation of Thomas White, James Wilson and Ebenezer Sibly. The last in particular, he thought, 'has done incalculable injury to this noble science.'[41] By now, he had won some acclaim among other astrologers, at least; the editors of *The Conjurer's Magazine* favoured him over the 'faulty and erroneous' work of Sibly,[42] and one C. E. Wynne of Portland Place, London, published an adulatory *Address to Mr. John Worsdale* (1816).

This initial success, however, faded out in the nineteenth century. Worsdale was a remarkable, and remarkably late, heir of the Ptolemaic reformers. He probably represents the last gasp of anti-scientific naturalism at any learned level. His kind of judicial astrology, uncompromisingly 'traditional' and technically highly demanding, continued to attract a small number of other judicial astrologers; but it held no promise of winning any wider appeal. It was too ostentatiously difficult to be popular; too religiously and intellectually unorthodox to achieve any mainstream acceptance (quite apart from the enduring stigma of astrology as such); and too narrowly internal and backward-looking for the Victorian middle-classes who might otherwise have been interested.

Yet judicial astrology succeeded in seeing out the eighteenth century with the promise of better days to come. To see how, we must turn to Worsdale's southern rival, Ebenezer Sibly (1751–1799).[43] Of humble origins, he had two brothers and a sister about whom we know nothing except in the case of his younger brother, Manoah (1757–1840). Manoah, like Ebenezer, was a capable scholar. He mastered many languages and first made a living as a bookseller. He later obtained a job in the Bank of England, and eventually rose to the position of a chancery office principal. In 1790, he became a Swedenborgian preacher. By this time, he had translated and published several astrological texts: Placidus's *Astronomy and Elementary Philosophy* (1789); *A Collection of Thirty Remarkable Nativities . . . from the Latin of Placidus de Titus* (1789); and Partridge's *Supplement to Placidus de Titus* (1790). In 1786, he also published a 'revised, corrected and improved' edition of John Whalley's 1701 translation of *Ptolemy's Quadripartite*. Thus this fundamental text was made available for yet another generation of English astrologers.

Ebenezer himself lived in Bristol (where he was born), London, Ipswich and eventually Portsmouth. He was a practising physician with an MD obtained from King's College, Aberdeen. His politics – like those of Worsdale and most active eighteenth century judicial astrologers – were radical but constitutional Whig, and he whole-heartedly supported the American Revolution. His metaphysical and intellectual commitments, in addition to astrology, included Freemasonry and animal magnetism. At the same time, he was a 'modern', who was well-read in the scientific literature of his day. In his books he therefore freely mixed Newton, Priestley and Lavoisier with Paracelsus, Hermes Trismegistus, and his astrological predecessors. This was no mere eclecticism; Sibly fervently believed that modern scientific knowledge, while valuable, needed to be supplemented by the vitalism and spirituality of the natural magic tradition. Only in this way, he felt, would it be possible to preserve the crucial connections between man, the microcosm and the universe, the macrocosm, and their equal dependence on divinity.

Sibly was a prolific writer. His *A New and Complete Illustration of the Celestial Science of Astrology* (1784–8) – sometimes appearing as *A New and Complete Illustration of the Occult Sciences* – was the first major public statement of astrology for many years. It appeared in four parts, and ran to over a thousand pages; by 1817, it had already gone through twelve editions, and continued to be reprinted until 1826. Nor was this Sibly's only work. He carried on the publishing tradition of popular astro-medical self-help established by Culpeper and Salmon, by editing a new edition of *Culpepers's English Physician and Complete Herbal* (1789); this work reached thirteen editions by 1812. He also found time to produce *A Key to Physic and the Occult Sciences* (1794, with five editions by 1814). And Sibly wrote another, purely astrological, book, *Uranoscopia, or the pure language of the Stars* (c.1785–7), with horoscope-blanks and tables.[44]

Sibly's *Illustration* was a comprehensive restatement of judicial astrological doctrine, and none the less so for being largely derivative of seventeenth-century material. He covered its genethliacal or natal, horary, political, and meteorological branches, interspersed with current knowledge and miscellaneous observations of interest. He also ventured some original prophecies. In volume three (published in 1787), he carried on the provocative astrological tradition of apparently anticipating spectacular events – this having largely languished since Lilly's and Edlin's forecasts of the last plague and the Great Fire of London. Two years before the French Revolution, Sibly analysed a figure of the Sun's ingress into Aries on 19 March, 1789. He concluded that

In fine, here is every prospect, from the disposition of the significators in this scheme, that some very important event will happen in the politics of France, such as may dethrone, or very nearly touch the life of, the king, and make victims of many great and illustrious men in church and state, preparatory to a revolution or change in the affairs of that empire, which will at once astonish and surprise the surrounding nations.[45]

It is unlikely that the general impression on readers of such a prediction, in the light of subsequent events, was erased by any number of accompanying but unsuccessful predictions to the same general end. In any case, Sibly himself, unlike some astrologers, was not particularly prone to making extravagant prophecies.

The *Illustration* shows clearly Sibly's enthusiastic eclecticism and omnivorous erudition. The consequences were important. His combination of experimental natural philosophy (including the latest medical and scientific advances) with occult and mystical knowledge resulted in a judicial astrology that, on the one hand, was defined naturalistically – 'the science which we call Astrology, is no more than the study or investigation of this frame or model of Nature, with all its admirable productions and effects.' On the other hand, his approach also emphasized astrology's magical elements: macrocosm and microcosm, sympathy and antipathy, the doctrine of signatures, and so on. Unlike Worsdale, for whom both new science and old magic were corruptions of the true astrology, Sibly embraced both, and asserted – in a confident reaffirmation of Lilly's democratic–populist astrology – that astrology 'is a science which all may attain to, by common diligence and application.'[46]

It is striking to find such a combination. To appreciate its significance, we need to recall that the principal division between judicial astrologers a century earlier had been between those inclined to magic and populism (like Lilly) and those advocating natural philosophy and elitism (like Gadbury). As open advocacy of the former became too dangerous, this split was then replaced by one between the latter position and the neo-Aristotelian naturalism of Partridge, both claiming the mantle of rational and elite natural philosophy. As we saw, magical astrology was banished to a plebeian ghetto. Sibly's astrological amalgam of science and magic – and its appreciative audience – was thus not simply paradoxical, nor even an aberrant throwback to the heady pre-consensus days of the 1640s and 1650s; it was something new.

So too was its success, as shown by his sales. The ready market for Sibly's astrology, on the eve of the nineteenth century, is further evidence of a growing middle-brow no-man's-land, formerly on the borders of popular magic but under the sway of elite religion and natural philosophy. To an unprecedented extent, an intermediate class of the 'semi-erudite'[47] now felt free to mix the two, in a fusion that was therefore neither strictly one nor the other. This development opened up new possibilities for legitimating the judicial astrological experience – that is, the experience of truth in the interpretation of a horoscope (the supposed validity of which we are not, for reasons already explained, interested in debating) – in terms of a popular magical science. Without having been eliminated (representatives exist today), the alternative strategies of Worsdale's rationalistic traditionalism and Henry Andrews's sober Christian pietism were reduced to being the choice of a relatively eccentric few.

In colonizing this new terrain for judicial astrology, Sibly led the way; succeeding Victorian astrologers, such as Zadkiel and Raphael, essentially followed the same pattern. With almanacs for readers who would have disdained *Moore's*, but who devoured the blend that Sibly pioneered – with daily forecasts soon thrown in for good measure – their relative success confirms that this judicial astrology had found an enduring middle-class audience. Still later peaks (in the 1930s, late 1960s and early 1970s) have been accompanied by lulls, at most.[48] Despite both patrician and plebeian bemusement, judicial astrology is clearly here to stay.

6

High Astrology: Disappearance

Astrology and the Natural Philosophers

A third kind of astrology still needs to be located and examined in eighteenth-century England: high, philosophical or cosmological astrology. As such a description implies, this use of astrological ideas was concerned with questions about the structure, functioning and governance of the universe. Since the latter includes the Earth, this concern often resulted in implications and predictions about earthly functioning and governance. Of course, the latter point overlapped somewhat with both judicial and popular astrologies. The intellectual difference was that judicial conclusions were based on the interpretation of traditional horoscopes, while high astrology ostensibly drew conclusions directly from current observational and theoretical astronomy. So too, in its way, did popular astrology – often with reference to the same comets, supernovae, and so on. But here the distinction was between a self-defined 'sacred' astrology, based on theology and natural philosophy, and a 'profane' astrology, based on popular religion and folk wisdom – superstition and ignorance, respectively, in the eyes of its high counterpart. It goes almost without saying that the social differences between adherents of high and popular astrologies were vast, with the gulf between both of them and judicial astrologers being less so only by virtue of the latter's social location 'midway' between the other two.

This kind of astrology brings us squarely into the terrain of both natural philosophy and theology (since in this period the two were virtually inseparable). Given received historical wisdom, it is still possible to think, that by 1700, the scientific revolution had killed off astrology altogether.[1] We have seen how far from being true that is, whether of popular or judicial astrology. Within its own domain, however, the failure of astrology's scientific reformers, amid the ambivalence or hostility of natural philosophers such as Flamsteed, Halley and Hooke, and the polemicist fringe of More and Sprat, was indeed followed by the gradual disappearance of astrology from this kind of discourse. Given the historical circumstances – the deep distrust of astrology and its practitioners by elite opinion after 1660, the efforts of the Royal Society under the aegis of the experimental method to secure a role in the conservative post-Restoration settlement, and the fact that their natural objects of study (in which the planets and stars took pride of place) brought them into direct conflict with astrology – the reaction of natural philosophers was understandably hostile.

Upon closer examination, however, the real process at work turns out not to have been an outright extinction so much as a transformation, whereby hitherto openly astrological ideas were redescribed in natural philosophical terms. In the late seventeenth and early eighteenth centuries, Newton and his colleagues in natural philosophy and divinity initiated work on cosmological theories – particularly those concerning comets – which continued to dominate natural philosophy for decades to come. There is no doubt that the origins of these ideas were partly astrological: they were based on the standard premise of specifiable, if as yet unquantified, influences from the celestial bodies on earthly and human life. In addition, speculations about the implications and effects of comets had been part of the astrologer's stock-in-trade for centuries. Eventually, as these recycled ideas were accepted in their new scientific terms, among natural philosophers and their constituency, extinction occurred – in the sense, real enough, that they were no longer widely or easily recognized as astrology.

This process has often been ignored or misunderstood by historians. Sometimes, in a general effort to preserve the purity of the antecedents of modern science, individuals have been sorted into categories ('scientific' vs 'superstitious') and even divided internally – allowing Newton's alchemy, for example, to be rendered harmless, since it was hermetically sealed from the Newton which matters; likewise his politics, and so on.[2] Kepler and Van Helmont, among others, have suffered like indignities. Whatever this practice leads to, it is not historical understanding, and we shall avoid it here.

In order to trace the disappearance of high astrology, we first need to return to the late seventeenth century. On 19 November 1672, John

Wilkins, a founding Fellow of the Royal Society and Bishop of Chester, died. His colleague Robert Hooke (1635–1702) recorded in his journal: 'Dyed about 9 in the morning of a suppression of Urine', followed by 'a conjunction of Saturn and Mars', and (in larger letters) 'Fatall Day'.[3] This was written in the space normally reserved for meteorological reports, graphically confirming the extent to which astrology still occupied an epistemological space in which the categories of natural and social had yet to be clearly separated – even for the natural philosophers who, in order to secure a place in the latter world as the proper interpreters of the former, were leading the struggle to do so. In public discourse, their priority was to redescribe the relevant astrology; but when it was private, as here (and as we saw with John Evelyn and John Dryden), they were quite willing to indulge in openly astrological reasoning.

The political and religious unrest in the period 1679–81 was accompanied by a spectacular comet in the winter of 1680–81. This comet was exploited rhetorically by various parties involved in the Popish Plot, notably Ezerel Tonge in his *The Northern Star*. Even John Tillotson, later Archbishop of Canterbury, recorded in a letter that 'the Comet hath appear'd here very plaine for several nights . . . What it portends God knows: the Marquess of Dorchester and my ld Coventry dyed soon after.'[4] A few years later, John Edwards's *Cometomantia* (1684) vehemently condemned astrology. But in the same work, he admitted the intimacy of natural and civil events. As an emissary from God, the comet of 1680 in particular was 'of universal influence, and many Nations shall share in the Revolution and Occurrances which it shall produce.'[5] The use of terms such as 'influence' and 'produce' suggests that this was a case of 'Astrology is dead; long live astrology!' That it was so little remarked upon as such at the time – except by astrologers, whose opinions, by definition, were excluded as unworthy of serious consideration – is a sign of the effectiveness with which astrological concepts were being appropriated for various purposes without the stigma of their original affiliation.

John Flamsteed (1646–1719) came closer to a wholesale rejection. We have discussed his ambivalence towards the work of Goad. However, he adopted a severe attitude to judicial astrology in his unpublished 'Ephemeris' of 1674. He attacked 'ye Vanity of Astrology and the practices of astrologers', in particular their almanacs, about which he complained that 'the Vulgar have esteemed them as the very oracles of

God.' Apart from this affront to an astronomer himself struggling for funds and recognition, why should the credence given to astrologers have been so objectionable? Flamsteed himself posed and answered that question:

> Of what ill consequences their predictions have beene, and how made use of in all commotions of the people against lawfull and established sovereignity, the historys of all insurrections, and our own sad experience, in the late Wars, will abundantly shew the considerate; how they have erred in their Judgements, the same experience will informe us.

As the last part of that sentence clearly shows, for Flamsteed the epistemological or 'scientific' shortcomings of astrology were twinned with its undesirable social consequences. That is why he objected in the same breath to astrologers' 'pernicious predictions of ye Weather and State affaires' and 'theire credit with the vulgar'. Similarly, his objections to astrologers' *ad hoc* justification of predictions and their countless mutual disagreements were not merely detached criticisms; they derived much of their force from the way such predilections paralleled and invoked the illuminationist antinomianism of the radical sectaries and their endless fission respectively.[6]

The following year, Flamsteed himself drew up a horoscope for the foundation of the Royal Greenwich Observatory – but only to scoff, it seems, since it was accompanied by the scribbled note, 'Risum teneatis, amici' (Can you help laughing, friends). The astrological habit of recording such moments must have lain deep however, or his ambivalence lasted long, for over twenty years later, on 30 June 1696, he attended the laying of the foundation stone for the Royal Greenwich Hospital. John Evelyn recorded that it took place 'after dinner at precisely 5 o'clock, Mr Flamsteed, the King's Astronomical Professor observing the punctual time by instruments.'[7]

In 1677, Flamsteed had expressed the hope that 'ye fearful predictions' of astrologers would be undermined if the periodicity of comets could be demonstrated.[8] But periodicity in itself constituted no threat to astrology; indeed, the periodicity of the great conjunctions was one of the very things that had always made them attractive and useful to astrologers in the past. As late as 1789, Henry Andrews, writing in *Moore's*, maintained that the regular return of comets in no way argued against their influence, since history revealed a cyclical pattern, and the same causes may be assumed to have the same effects. Indeed, the work of

Newton himself, along with other natural philosophers in this period, shows that periodicity in itself did nothing to eliminate the eschatalogical, religious and prophetic functions attributed to comets.

Newton, the Occult and Astrology

Modern astrologers and their textbooks tend to offer sops to history that oddly resemble the historical myopia of most scientists. Perhaps the most common is the story of an encounter between Newton and Halley, in which the latter makes bold to disparage astrology. Newton replies to the contrary, asserting loftily that 'I have studied the matter, Sir, and you have not.' Astrologers' attachment to this comforting anecdote is strong, successfully holding out against a complete lack of any supporting historical evidence. According to Conduitt, Newton's initial interest in mathematics was piqued by an astrological book he purchased at Stourbridge Fair as a young man, in 1663. But that is a slender reed, and an unlikelier hero and defender of astrology than Isaac Newton (1642–1727) is difficult to imagine.[9]

In the 1680s, Newton established to his own satisfaction the elliptical orbit of comets. In the 1690s, he was joined by Edmond Halley, and by 1696, they had worked out 'at least two closed and periodic cometary paths'. They continued to develop and publish this work into the 1720s. These are the bare historical bones, but, as Simon Schaffer has commented, 'the real significance of this research programme cannot be understood within such an isolated context.' Vital motivation (at least in Newton's case), and the wider uses to which this natural philosophical work was put, were informed by a deep commitment to two sets of ideas. Both are marked out by the putative status of comets as officers of divine power. Firstly, comets were held to be responsible for maintaining the stability and vitality of the cosmos, which would otherwise degenerate and expire; their tails in particular were held to replenish the vegetative life on Earth. Secondly, part of the sacred power of comets in Newtonian natural philosophy was to sweep away the corrupt idolatrous and polytheistic vestiges of paganism – specifically including 'astrologers, augurs, auruspicers &c . . . [and] such as pretend to ye art of divining'. Thus 'scientifically' and theologically construed, a fundamentally astrological concept of comets could safely be used even as a rhetorical weapon against astrologers.[10]

It should also be noted that in Newton's private alchemical work – the influence of which on his theory of matter is well known, if its extent is controversial[11] – he was quite happy to use astrological concepts and reasoning. For example, he held that the best water for such work was drawn, in his own words, 'by the power of our sulphur which lies hid in Antimony. For Antimony was called Aries by the Ancients. Because Aries is the first Zodiac Sign in which the Sun begins to be exalted and Gold is exalted most of all in Antimony'.[12] Here we see once more the familiar disparity between contemporary natural philosophers' private and public attitudes towards astrology. This was complicated in Newton's case by his profound commitment to ancient Hermetic doctrines, which included astrology, combined with an equally deep dislike of what he viewed as their corrupt modern descendents, including astrologers.

It has also been suggested that astrology provided Newton with the basic concept of gravity, namely, universal action-at-a-distance.[13] That is probably taking the idea too far, although it would fit with Newton's plundering of astrological cometography for his own natural philosophical purposes. What is certain is that given his place in early modern natural philosophy, and the subsequent development of science, including his theory of universal gravitation, Newton had a disproportionately large impact on the astrology of his time and since. That impact centred on his theory's redefinition of the idea of 'occult'. As Keith Hutchison has shown, Newton split occult causes into those which are acceptable and those which are not. Only the former – of which those causes generating gravity are an example – are universal, and permit reliable empirical detection of their effects.[14] In both respects then, astrology was damned: in its attribution of different qualities or principles unique to each of the planets, and in its empirical unreliability.

With the much-hailed success of his theory, Newton further enshrined the experimental method and its criteria for truth which proved such an insuperable obstacle for the 'scientific' reformers of astrology. Stranded on the wrong side of that barrier, astrology increasingly became a thoroughly occult knowledge, in the modern sense of the word: that is, an esoteric, recondite, mystical pursuit. Thus modern astrology, in relation to modern science, is not so much an inexplicable survivor of pre-scientific ignorance and superstition as a scion of the scientific enlightenment itself. In the meantime, these intellectual consequences were spread and cemented by Newton's huge reputation throughout the eighteenth century. Thus magic (that is, the occult in the discredited sense) was further excluded from polite discourse. This remained highly effective until the middle classes (with deplorable buying power and lack of discrimination) began, in the 1790s, to mix things up again. In the meantime, the strategic disappearance of astrology that Newton pioneered was seen through by a host of colleagues and commentators.

One was David Gregory (1661–1708), a pupil of Newton and Savilian Professor of Astronomy at Oxford. Gregory's principal textbook of 1702 included an essay by Newton on the Hermetic 'ancient philosophy', a major source of ideas for Newton on comets as quasi-planets with closed orbits.[15] Its supposed originators, the 'Chaldeans', had also traditionally been regarded (with some justification) as the fount of astrology; their very name was synonymous with 'astrologers'. So it may be surmised that Gregory agreed with Newton on the importance of this fundamental source. Yet in an unpublished manuscript written in 1686, he vehemently denounced astrology as fraudulent and seditious. Like Flamsteed, he objected on grounds that were at once epistemological and political, 'scientific' and social:

> it is hardly believable what great damage those predictions invented by Astrologers and ascribed to the stars have caused to our best kings, Charles I and II . . . Therefore we prohibit Astrology to take a place in our Astronomy, since it is supported by no solid fundament, but stands on the utterly ridiculous opinions of certain people, opinions that are so framed as to promote the attempts of men tending to form factions.[16]

The link between the epistemological and political problems, as perceived by men such as Gregory, was this: if astrology was a purely plastic subject, with no controls exerted by 'Nature', then it could be endlessly misused for extreme ends by irresponsible interpreters, that is, those with no stake in maintaining the existing social order. Whereas if people could be convinced that the prophecies of natural philosophers and astronomers were (by contrast) ineluctably drawn from Nature – in other words, that they were 'objective' – and therefore sanctioned by God, the architect of Nature, then all reasonable people would perforce be obliged to concur. Of course, the necessary interpretation could only be done by those qualified to do so. And the social process of such qualification could generally be relied on to weed out those who were unsound.[17]

Gregory's undilutedly Jacobite sentiments (probably the reason why this lecture was never published) were the very opposite of Newton's, who abhorred Catholicism and had risked his career by opposing James's appointment at Cambridge. But the agreement between them on the corruptness and despicability of astrology, and on the status of comets as agents of divine will, nicely emphasizes two things: firstly, the completeness of astrology's exclusion from polite and educated discourse in the late seventeenth and early eighteenth centuries. That in turn points to the social common ground underlying the divisions between Whig and Tory, High and Low Church. I am not suggesting that such divisions were not real or important; only that the breach between patrician and plebeian ran deeper. All parties of the former, whatever their particular persuasion, were agreed that (in the words of a 1721 manual of elevating thoughts) 'Those Men who are inclin'd to Atheism are most addicted to Judicial Astrology, because the Mind of Man must have something to rest upon; and when the solid Principles of Religion are thrown aside, it seeks a deceitful Refuge'.[18] Secondly, we can see that this exclusion included the adoption by natural philosophers of certain vitally important and useful astrological ideas, but without the former label and its associations. The disappearance of cosmological astrology was thus accomplished through a re-articulation in which the original ideas were rendered useful, since as part of that process, their only qualified interpreters were established as now no longer astrologers, prophets of the 'superstitious Vulgar' and 'the Mobb', but Christian natural philosophers.

Authorized Prophets

High astrological ideas, in their new incarnation, were not confined to an esoteric circle among the elite. Through eagerly received books and public lectures, they were disseminated and discussed among the educated and curious throughout the eighteenth century. This was accomplished by professional astronomers – witheringly but accurately described by J. H. Lambert in 1760 as 'authorized prophets' – and their interpreters.[19] Some of this carried over from the late seventeenth century, as with John Ray's *Miscellaneous Discourses concerning the Dissolution and Changes of the World* (third edition, 1713). Ray was responding to a widespread concern with 'the fates of Kingdoms and Commonwealth, especially the Periodic Mutations, and final Catastrophe of the World' – as good a definition as any of one of astrologers' principal stamping grounds.[20] But the principal vehicles for this process were Newton's Latitudinarian and natural philosophical acolytes, Clarke and Whiston. Samuel Clarke (1675–1729) was an influential popularizer of Newton's ideas (through, for example, his Boyle lectures of 1704–5). In a sermon published in 1730, he stressed that 'those always who have least knowledge of God, and least Trust in his Providence, and least Understanding in the true System and Powers of Nature, have the greatest Confidence in groundless Pretences and unwarrantable Methods of pursuing Knowledge'. Specifically, 'to pretend to know things by the Stars, which introduces Fatality and destroys Religion, is not much different from pretending to know them by Arts that have worse names.'[21]

William Whiston (1667–1752) was another Boyle lecturer (in 1707) and a friend of both Clarke and Newton, whom he briefly succeeded as Lucasian Professor. Borrowing an idea first mooted by Halley – that a comet may have been responsible for the biblical Deluge – Whiston joined it with Newton's general views on comets. This provided the basis for his influential lectures and books, widely debated in coffee-house circles: *A New Theory of the Earth* (1696), *A Vindication of the New Theory of the Earth* (1698), and *Astronomical Principles of Religion, Natural and Revealed* (1717, reprinted in 1725). In Whiston's system, not only had a comet precipitated the Deluge, but the revelation of Newtonian cometography presaged what will be a comet's final act: the destruction by fire and divine restoration of the Earth to its pre-lapsarian purity. In his own words, comets 'cause vast Mutations in the Planets, particularly in bringing on them Deluges and Conflagrations . . . and so seem capable of being Instruments of Divine Vengeance upon the wicked Inhabitants of any of those Worlds; and of burning up, or perhaps, of purging, the outward Regions of them in order to [effect] a Revolution.'[22] What astrologer has ever outdone this in apocalyptic abandon? Whiston's ideas were even appreciatively received by the same kind of people who had formerly followed Lilly's every word. Along with Flamsteed and Halley, he was identified as an astrologer by a popular publication in 1715. A couple of decades later, 'No little consternation was created in London in 1736 by the prophecy of the famous Whiston, that the world would be destroyed in that year, on the thirteenth of October. Crowds of people went out on the appointed day to Islington, Hampstead, and the fields intervening, to see the destruction of London, which was to be the "beginning of the end".'[23]

The erstwhile astrology is plainly present in Whiston's discourse, including an ingenious suggestion as to its origins. Nor was it restricted to comets:

> In our Accounts of the Deluge and Conflagration, there is a notable conjunction of the Heavenly Bodies indeed; not such an Imaginary one as the Astrologers so ridiculously make a stir about . . . but a real one with a Witness; when three of the Heavenly Bodies, the Earth, the Moon, and the Comet, not only are in an Astrological Heliocentrick Conjunction . . . but are really so near as to have the mightyest effects and Influences on one another possible . . .
> Corollary: 'Tis not improbable but the ancient Tradition, that the Deluge and Conflagration some way depended on certain remarkable Conjunctions of the Heavenly Bodies, mis-understood, and afterward precariously and widely mis-apply'd, might give occasion and rise to Astrology; or that mighty quoil and pother so many in all Ages have made about the Conjunctions, Oppositions, and Aspects of the Heavenly Bodies, and the Judiciary Predictions therefrom; which even the Improvements of solid Philosophy in our Age have not been able yet to banish wholly from among us. [24]

The High Church and Tory opponents of Newtonianism viewed it as readmitting occultism by the back door, hidden in the concept of gravity. Rather like many of the common people, they too were unconvinced by disclaimers to the contrary. Eventually, Whiston sailed too close to the wind. His open indulgence in unorthodox speculation, especially Arianism, placed him beyond the pale, and he was reluctantly abandoned by Newton. Clarke's career too suffered for the same reason. But this was a dispute internal to the dominant elite, between those who (in this context) wanted a limited toleration for occult forces, scientifically reborn, and those who wanted none. The vulnerability of the Newtonians to accusations of heterodoxy and occultism undoubtedly spurred them to specifically distance themselves from astrology, which in any case they saw as little better than papist idolatry. Their conservative opponents, on the other hand – along with Tory natural philosophers, such as Halley and Gregory – concentrated on the contribution of astrologers to bringing about the Civil War and the death of Charles I. But the social basis for both charges was essentially the same; and so were the consequences for high astrology.

From the early eighteenth century, there was a boom in the publishing of 'popular' natural philosophy: versions of Newton's works, textbooks, encyclopaedias, and scientific literature for women and children. [25] These publications were aimed at an increasingly wide audience, extending throughout and even beyond the educated middle classes into coffee-houses and 'penny universities'. Whiston was not the only prominent lecturer and writer in this area. Another was William Derham (1657–1735), one of the annual lecturers underwritten by funds left by Robert Boyle. In his immensely successful *Astro-Theology: Or a Demonstration of the Being and Attributes of God, from a Survey of the Heavens* (1715), Derham managed to discuss that subject without a single reference (even disparaging) to astrology – a sign of the latter's weakness, by now, in higher intellectual and social circles. In strikingly astrological terms, however, Derham opined that 'the great Creator hath made the Moon to be of admirable use to our Earth. And so wisely hath he conceived his Works, that they are mutually serviceable to one another . . . Thus . . . [the Moon] perhaps makes such like returns of Influx as I said the earth receives from her. For it is not to be doubted . . . but that there is a mutual intercourse and return of their Influences and good Offices'.[26]

Another Boyle lecturer and Fellow of the Royal Society, John Harris (?1666–1719), dismissed astrology in a single sentence in his *Lexicon Technicum* (1704). He added candidly that 'as I wish that such a ridiculous piece of Foolery as this may be quite forgotten, so I have every where omitted explaining any of its Terms, unless they fall within Astronomy.' The entry under 'Infection', however, explains that 'Mr. Boyle, is inclin'd to believe (tho' he had no Opinion of Judicial Astrology in other respects) that the Planets may have some Physical Influence or Operation on Bodies belonging to our Globe'. Harris went on to elaborate the idea that each planet possesses 'its own proper Light distinct from every other, which Light . . . must be accompanied with some peculiar Tincture, Virtue, or Power.'[27] This had been the chief theoretical justification for judicial astrology for centuries. Now it was being salvaged and divided between medicine and astronomy.

Turning to the popular and influential *Cyclopaedia* (first edition, 1728), Ephraim Chambers (1680–1740) also made the customary sharp distinction between natural and judicial kinds of astrology. 'Influence' he mockingly defined as 'a Quality supposed to flow from the Bodies of the Stars, or the Effect of their Heat and Light, to which the Astrologers vainly attribute all the Events that happen on Earth.' Yet cosmological natural astrology, incorporated into natural philosophy, received a full and remarkable justification. Citing Boyle's *History of the Air*, Chambers argued that 'the Course, Motion, Position &c. of the heavenly Bodies' govern all meteorological phenomena. Furthermore, "tis also clear that every Planet must have its own proper Light, distinct from that of any other; Light not being a bare visible Quality, but embued with its specifick Power'. That is, the light of the Sun, when reflected from each planet, 'must share or receive somewhat of the Tincture thereof'; whence according to the angle between the planet, the Sun, and the Earth – in other words, astrological aspects – and their respective positions and distances, 'the Powers, Effects, or Tincture, proper to each, must be transmitted hitherto, and have a greater or lesser effect on sublunary things.'[28]

Chambers's *Cyclopaedia* was a model for Diderot and D'Alembert's famous *Encyclopédie* (1751–65), which began life as a translation of it. These arch-Enlightenment intellectuals accepted and reproduced the same schema as Chambers, permitting each planet 'un pouvoir spécifique' which affects, via aspects, 'les êtres sublunaires'. Natural astrology is therefore 'une branche de la Physique ou Philosophie naturelle', and 'M. Boyle a eu raison quand il a fait l'apologie de cette Astrologie dans son *Histoire de l'Air*'. Protected by the same sterilized version of high astrology, and therefore with the same lack of any sense of inconsistency, these authors denigrated its judicial and popular branches.[29]

On the face of it, it is difficult to imagine what more judicial astrologers could have asked for by way of intellectual rationale or justification, and this in authoritative reviews of contemporary knowledge. Given the difficulty of drawing any firm line between natural and judicial astrology, just such a set of ideas had supplied them with an intellectual haven for centuries – to wit, judicial astrology may be abused by irresponsible practitioners, but it possesses none the less a firm basis in natural philosophy and theology. But to view early modern natural philosophers as merely hypocritical, extending the same rationale into the eighteenth century, would be to miss the point. On the contrary, we are seeing here the end of high astrology, simply (but quite effectively) because the latter was no longer recognizable *as* astrology, its formal identity notwithstanding. 'Astrology' was now *just* popular or judicial.

By 1771, the *Encyclopaedia Britannica* gave astrology as such exactly five lines: 'a conjectural science, which teaches to judge of the effects and influences of the stars, and to foretell future events by the situation and different aspects of the heavenly bodies. This science has long ago become a just subject of contempt and ridicule.'[30] This entry compares with sixty-six pages on astronomy – a section which contains, however, considerable speculation on the cosmological functions and effects of comets, including the probable presence thereon of sentient beings.[31] This idea, a constant theme in eighteenth-century astronomy, was pursued with an intellectual abandon which makes much astrology look comparatively tame. Not having the latter's associations, however, it flourished. James Ferguson (1710–76) was a leading popular scientific writer, producing widely selling books such as *Astronomy Explained on Sir Isaac Newton's Principles* (1756), and *The Young Gentleman's and Lady's Astronomy* (1768). In 1754, he speculated that comets

> are not destitute of Beings capable of contemplating with Wonder, and acknowledging with gratitude, the Beauty, Wisdom, and symmetry of the Creation . . . If farther speculation is permitted, may one not suppose they are peopled with guilty Creatures reclaimable by Sufferings, as we are on the Earth; and, like every thing else that falls under our Observation, may be subservient to other secondary Purposes; such as recruiting the expended Fuel of the Sun; supplying the exhausted Moisture of the Planets; causing Deluges and Conflagrations for the Correction and Punishment of Vice?[32]

Richard Turner (?1724–91), in *A View of the Heavens* (1765), also conjectured that comets are 'habitations of the damned', but added that 'others have concluded them, with greater Probability, to be the Executioners of God's vengeance on sinful Worlds; by scattering their Baneful Influences on the Inhabitants, or dashing the Planet to Pieces, and reducing it to its chaotic State again.'[33]

The reiteration of the ancient argument that the stars and planets are secondary causes, acting as instruments of God; the view of comets as harbingers of terror and divine retribution; the ideas of cosmic animism; 'Influences'; the old astrology in all this is plain enough. Set in a modern, 'scientific' context, however, it was now safely separated from that art which – in the words of John Hill (?1716–75), author of *Urania: or, A Compleat View of the Heavens* (1754) – 'lost its credit, while enthusiasm took the place of rules'. Most scientific writers were content to reiterate the charges of 'idolatry and superstition', but Hill went further:

> we are not to doubt but that the same force of imagination, which leads one man to imagine he is a tea-pot, or dish of meat . . . should be able to persuade another that he has inward light and supernatural notices of events; or that brooding over a set of idle schemes and unmeaning figures, he shall fancy he is able to see into futurity . . . this disorder is a degree of madness.'[34]

The familiar tones of gentry hegemony are clear enough, although the explicit imputation of psychological disorder is an interesting development. (Hogarth's portrayal of enthusiasm, published in 1762, made the same general point about Methodism.)

A similar process was taking place in relation to astrological physic. In the entry on medicine in the *Encyclopaedia Britannica* (1771), epileptic fits were reported to occur 'according to the quadratures of the Moon, but especially about the full or new Moon.'[35] The authority cited was not Culpeper, Blagrave or Salmon, but the most eminent physician in early eighteenth century London, Richard Mead (1673–1754). Mead's patients included Newton and Halley, and he was himself a Vice-President of the Royal Society in 1717. (He also briefly taught William Stukeley medicine.)[36] In 1704, Mead published *De Imperio Solis ac Lunae in Corpora Humana, et Morbis inde Oriundis*. This was a widely read work; besides appearing in several editions of his collected works and other collections, it was reprinted in 1710 and 1746, and appeared in English in editions of 1708, 1712, and 1748. Later in the century, Mead's ideas reappeared in Mesmer's MD dissertation, *De planetarum influxu*, which discussed planetary effects on rhythms of health and disease.[37]

Mead argued that the gravitational interactions of the planets (especially the Sun and Moon) affected the Earth's atmosphere with significant physical and aetiological consequences. This explained such phenomena as the periodicity of menstruation and epilepsy in relation to the Moon. Mead also allowed for the effects of 'Powers of Comets and Planets' (other than the Sun and Moon). He even cited Goad's *Astro-Meteorologia*, for readers seeking 'a fuller account.'[38] Mead, the contemporary exemplar of learned medicine, was clearly willing to go farther towards explicitly acknowledging astrology, when drawing on its explanatory resources, than were astronomers. The reason probably lies in astronomy's greater development as a 'hard' science, with a relatively stable set of objects to study. The planets, stars and comets construed as 'purely' natural objects could then be used to underwrite more sweeping claims to objectivity – claims which both permitted and required a more thorough assimilation (and sharper explicit rejection) of astrology, with its putatively subjective and irrational handling of the same objects, than in medicine, where such a taint could not be so readily banished.

It is not necessary to trace here the further course of natural philosophy *vis-à-vis* astrology. In general, the disappearance of high astrology continued, to the point of completion, as part of the increasing professionalization of astronomy in the nineteenth century. Amid the recondite theoretical, mathematical and observational developments, and lacking any serious rival articulation, high astrology ceased to exist. (Although bearing in mind previously exaggerated reports of death, we may not say its own was necessarily complete or permanent.) In the wake of this process, the way was left open for others, working in less-developed scientific fields, to continue to engage in quasi-astrological speculation with impunity. In 1799, Humphry Davy (chemist, philosopher, and future President of the Royal Society) speculated on the presence of light in the arterial blood, in the form of an etherial fluid which maintains the sentience of the brain. He suggested, in terms whose resonances are unmistakable, that 'We may consider the Sun and the fixed stars, the suns of other worlds, as immense reservoirs of light destined by the great ORGANIZER to diffuse over the universe organization and animation.'[39] Many of Davy's contemporaries – from the readers of Ebenezer Sibly to those of Old Moore – would have agreed with such speculation in terms with which he might have felt less than comfortable. Who can say that their understanding, however different, was inferior to his?

7

The Reform of Prophecy

Patricians, Plebeians and the Middling Sort

In the preceding chapters, I discussed the complex fate of early modern English astrology. We now need to accept the challenge of these changes, and sharpen the basic ideas that have made them initially comprehensible. Their fundamental historical context is what Peter Burke has identified as 'the reform of popular culture' – that is, the withdrawal of the governing and educated elite from the social and cultural world of the great mass of people, in which it had formerly felt free to participate.[1] This was accompanied by a determined and systematic attempt by some of the elite – especially the militant clergies of the Reformation and Counter-Reformation in the sixteenth century – to reform popular values and beliefs. Scrutiny for heresy, idolatry and superstition intensified, as ecclesiastical supervision tightened; and this was paralleled by increasingly centralized political control.[2] Burke also identified a later phase overlapping with the first, beginning about the middle of the seventeenth century, in which the laity took the initiative, and the emphasis shifted from 'saving' the people to 'improving' them.

This process was European. The rate and extent of change varied within national (and local) boundaries, but throughout Western Europe various practices hitherto tolerated by the Church and intelligentsia were attacked and driven underground. In concentrating on astrology in England, I have necessarily sacrificed a broad comparative picture, both nationally and respecting other stigmatized practices; but a word on France at least seems called for. At first glance the situation seems very similar to that of England. Beyond the shared 'decline' of astrology amid the basic trends just noted, however, there lay a different set of causes and effects. Considerably greater centralized control, both political and religious, meant that when French elite opinion turned against judicial astrology in the second half of the seventeenth century – as marked, for example, by Colbert's decision to exclude it from the Académie des Sciences in 1666 – its essential court patronage dried up, and there was little else to fall back on. Natural astrology was largely appropriated by natural philosophy, much as it was in England, but popular astrology retreated to a base of stereotyped and repetitive almanacs which lacked (as Capp has noted) the interest of the controversial English-style prognostication. This retreat was accelerated by the Enlightenment *philosophes*. In the late eighteenth and early nineteenth centuries, astrology reappeared as a legitimate subject of academic (particularly historical) inquiry. But compared with England, there was no real tradition of a middling judicial astrology to fructify the popular and philosophical kinds, or to sustain a later revival, such as occurred in middle-class Victorian society.[3]

There is no doubt that the English evidence confirms the truth of Burke's general analysis, but it also demands greater specificity. In early modern England, the reform of popular culture occurred principally in the form of a new and momentous split between patrician and plebeian culture, which appeared after the Restoration of 1660. It divided neither the wealthy and the poor, nor the aristocracy and commoners, but the respectable, or 'better sort', on the one hand and the great mass of labouring people, or the 'vulgar', on the other. The former therefore included not only the nobility and gentry, but most of the nascent middle classes (a fact whose implications we will explore in a moment). Partly its result, this division then became, in turn, a major impetus for attempts to reform popular culture, customs, and beliefs. There was a fresh wave of such attempts in the final decades of the eighteenth century, which coincided with the closing of the ranks among the gentry between old Whig and old Tory.[4]

It was E. P. Thompson who first recognized the importance of this development, and outlined its momentous social and cultural consequences.[5] He further conjectured that the professional and commercial middle classes whole-heartedly adopted the essential values and ideas of their social superiors, remaining their eager clients until at least the 1770s, showing no determined signs of independence until the 1790s, and acquiring real power only in the 1820s. Hence their speaking in 'the hegemonic voice of the gentry', a view recently echoed by J. C. D. Clark.[6]

In fact, despite their enormous political differences, Clark's more recent study confirms a great deal of Thompson's analysis: the patrician–plebeian split, the assimilation of the middle classes by the former, and the importance of ideology. Clark demotes the unrest of the 1790s to a mere disturbance, and stresses the determinate force in eighteenth-century England of an Anglican clergy and intelligentsia, an elite aristocratic ethic, and a dynastic monarchy; but his principal difference with Thompson is that where the latter decries that force, he celebrates it.[7]

It could be argued that Clark goes too far, and that the national polity in the eighteenth century was more diverse and unstable than he allows.[8] Admittedly, the power of Church, King and gentry (and the often reactionary conservatism of their views) is indisputable. In addition to internal political and religious differences, however, the Glorious Revolution in 1688–9 was followed by increasing concessions by the traditional monarchical, aristocratic and ecclesiastical elites to metropolitan professionals: merchants, manufacturers and retailers, public servants, journalists and authors, lawyers, educators, and architects.[9] True, the latter avidly sought to emulate the former; but an alliance of interests is not the same thing as an identity. For that reason, it is the middle classes – somewhat neglected to date, compared to their purely patrician and plebeian peers – who lend much of the interest to eighteenth-century social history. Chronically restless and unstable, increasingly more entrepreneurial and less deferential, they showed unmistakable signs of dissent in the 1770s and 1780s. But the resulting agitation for reform was crushed by Pitt's White Terror in the late 1790s, in the context of a general reaction against the excesses of the French Revolution.[10]

Meanwhile, the transformation of the popular classes into plebeian continued apace. Throughout the 1780s and 1790s, aristocratic wealth and estates grew at the expense of small and intermediate holdings; enclosures ate up public commons and rights alike, and the distance grew between the rural middle class and agricultural workers. As Cobbett warned, 'When farmers become gentlemen their labourers become slaves.' Thus a working class, increasingly self-conscious (and therefore potentially self-directed) came into being.[11]

Although this was a social process, it is not possible (as even such a brief account makes clear) simply to leave out politics – not least because of the extent to which the initial withdrawal of the gentry was 'hastened by a fearful reaction to the signs between 1640 and 1660 that the world

155

might be turned upside down', and maintained by subsequent events that rang the same alarm bells.[12] The resulting reaction against 'enthusiasm' continued to stimulate patrician solidarity, at least as late as the crushing of Jacobinism at the end of the century.

The history of astrology in the context of these events confirms, and fills out, E. P. Thompson's prescient insight that surviving magical beliefs in eighteenth-century England did not constitute a random muddle of popular ignorance, but 'a rather more coherent mental universe of symbolism informing practice'. That coherence, he suggested, 'arises less from any inherent cognitive structure than from the particular field of force and sociological oppositions peculiar to eighteenth century society; to be blunt, the discrete and fragmented patterns of thought become integrated by *class*.'[13]

Thompson has characterized this period as 'class struggle without class', that is, without the mature classes which emerged in the nineteenth century industrial revolution.[14] Without claiming that social class is always a dominant (or even relevant) historical consideration, it has clearly proved indispensable to any understanding of early modern astrology – a field, in other words, where one might well have had doubts as to its relevance. Taken with other historical work, I can therefore see no justification for refusing social class its due weight in pre-industrial history.[15] Of course, there are limits and qualifications: the 'horizontal' ties of community, for example. And it is only for contingent historical reasons that considerations of gender, race and nationality do not also figure more importantly here. But these surely interact with the effects of class, in various complex ways, rather than obliterate them.[16]

I am not arguing for a reductionist concept of classes (let alone culture) which ties them closely to the mode of production, and pretends to derive their character directly from their place therein. Such limpid simplicity may be desirable in a work of philosophy, but it is a fairy-tale in one of history or the human sciences.[17] Nor will production's 'determination in the last instance' do – a nostrum with no cash value except to Marxist scholastics.[18] This is not to diminish the importance of economic and material considerations. Rather, as Paul Veyne put it, 'tout est historique, tout dépend de tout (et non pas des seuls rapports de production), rien n'existe transhistoriquement et expliquer un prétendu objet consiste à montrer de quel contexte historique il dépend.'[19] But in my view, and in the light of this research, the following passage still puts the matter best:

class is not, as some sociologists would have it, a static category – so many people standing in this or that relation to the means of production – which can be measured in positivist or quantitative terms. Class, in the Marxist tradition, is (or ought to be) *a historical category*, describing *people in relationship over time*, and the ways in which they become conscious of their relations, separate, unite, enter into struggle, form institutions and transmit values in class ways. Hence class is an 'economic' and it is also a 'cultural' formation: it is impossible to give any theoretical priority to one aspect over the other.[20]

It is quite possible to have doubts and misgivings about Marxism, while recognizing its unique contribution in such a formulation.[21]

Necessary refinements and additions notwithstanding, this approach has now shown its worth in the previously befogged history of astrology. Already stigmatized by Protestant divines (but without great social force or consistency) as purveyors of superstition, English judicial astrologers paid dearly for their moment of glory during the Interregnum. Astrology *en tout* was caught up in the ensuing wave of elite revulsion (and to some extent, popular exhaustion) against enthusiasm. Efforts by judicial astrologers to escape its effects by reforming astrology into a rational natural philosophy (whether dominant, allied with the Royal Society, or dissident, as radical Whigs) failed. Unchecked, those effects extended far beyond the persecution of astrologers as dangerously irresponsible prophets, the censorship of their almanacs as 'oracles to the vulgar', and the diatribes of divines, natural philosophers and men of letters. The genteel identification of astrology as enthusiastic, and therefore (like enthusiasm itself) vulgar, became fixed in the minds not just of a few authorities but of an entire social class – a development only made possible by the unprecedented degree of patrician withdrawal and self-consciousness after 1660. Astrology soon became a touchstone, a marker beyond which bitter political and religious differences became internal, with Newton in agreement with Gregory, and Swift at one with Addison. As we saw, it survived in the eighteenth century only beyond the pale (albeit an enormous area) as a part of plebeian life and thought – or more tenuously, as what remained of judicial astrology, on its provincial fringes. Within polite society, astrological ideas were untouchable until they had been reconstructed and renamed. In all, the hegemonic power of the gentry within the magic circle of the upper and middle classes stands confirmed; but so too, by the indisputable survival of popular astrology, do the limits of that power.

That situation remained relatively stable until the 1790s, when – precisely in line with other expressions of independence – we find a resurgence of judicial astrology among an urban and literate middle-class audience. (The congruence even obtains down to the 1830s, when that interest properly took hold with the new almanacs of Raphael and Zadkiel.) Not that the gentry's opinion of astrology had changed: the nineteenth century also saw a new wave of attacks by reformers and scientific writers, and prosecutions of astrologers. Meanwhile, the despised but undaunted readership of *Moore's Almanack* continued slowly to grow, defining the contemporary distinction between elite and vulgar. As this account suggests, in various ways and on both sides of that border, there seems to have been something like a class mentality at work. It now deserves a closer look.

Mentalities and Ideologies

The history of mentalities has been principally developed by a group of French historians. Braving the murky depths between the history of ideas, social history and psychohistory, they have taken on the histories of attitudes to death, sexuality, childhood and religious belief.[22] Not surprisingly, this has attracted considerable attention, including some apposite criticism.[23] The problem is that the emphasis on homogenous 'collective mentalities' – tacit, widely shared and long-lasting – tends to obscure real differences, whether over time or at any given time. It also results, as one historian put it, in 'the problem on which all history of mentalities stumbles, that of the reasons for and modalities of the passage from one system to another.'[24] Recognizing such problems, Jacques Le Goff has suggested that 'Class mentalities exist side by side with unifying mentalities, and their workings have yet to be studied.'[25] Yet this approaches a self-contradiction, if mentalities are supposed by definition to be global.

The history of astrology in early modern England has much to contribute here. *Prima facie*, we are surely dealing with a mentality of some kind. In terms of the basic or formal ideas, astrology's longevity and universality has few rivals. Many of its variations imply the same point, such as the analogical association of seven planets with seven metals, ages and so on; or the quasi-astrological sense of connectedness with the cosmos – whether through sympathy or influence – such that 'if someone tried to convince you that your liver is really somewhere far out in space and that it carries out its functions by mysteriously raying its products to you across the empty miles, your incredulous reaction would not be very different from that of a fifteenth-century astrologer who had just been told that the planet Mercury is located millions of miles away in space and is not organically connected to him, nor in any way a part of him.'[26] (Such a perception suggests, incidentally, that when the readers of popular almanacs learned that the Moon was in 'the Head', that is, the sign of Aries, it may have carried more literal meaning than we can readily appreciate.) Not that the astrological mentality was purely idealistic; as a seamless blend of cosmology, popular religion, physic, husbandry and the like, it held practical and concrete implications for almost every kind of situation.

Yet even with such a perfect test-case, what is striking is astrology's dramatic, if selective, change over time (a very short period of time, historically speaking) and the vast differences in its meaning for different groups that this entailed. The elite rejection of astrology – socially embodied, but initially stimulated by the political and religious events of 1642–60 – confirms that Michel Vovelle was right recently to welcome 'the return of the event' to the history of mentalities.[27] Above all, it shows the need to historicize and contextualize mentalities, principally by relating them to ideology. For that rejection was above all ideological, that is, socially and consciously interested.

I am not suggesting that ideologies arise in some direct way out of 'real' material or social interests, which in turn produce purely symbolic mentalities as epiphenomena. Certainly the latter embody such interests, as I have shown in the case of astrology – not only in the dynamics of its 'decline', but right down to such fine points as the ideologies of competing schools of reform, and the differences of astrological philosophy between Lilly and Ashmole. There is no longer any excuse for considering mentalities idealistically, without reference to such interests.[28] In its complexity and contingency, however, the same history also rules out any simple-minded reduction of the former to the latter.[29] As Chartier has argued, 'The representations of the social world are themselves the constituents of social reality.'[30] The only acceptable definition of 'ideological' – and one which perfectly suits the requirements of this history – is therefore something like those activities and situations 'in which there is a consciousness of contested representations of the world in play, in which social action takes the form of more or less explicit attempts to order or reorder the world.'[31]

At the same time, it is striking to observe the way in which a sharply ideological awareness of astrology gradually subsided, and a less conscious, more automatic, class mentality towards it arose. From the perorations of Samuel Parker ('the wildest and most Enthusiasticke Fanaticisme') and Thomas Sprat ('this frightful, this Astrological Humour') in 1666–7 to that of Swift ('all the wise and learned . . . do unanimously agree to laugh at and despise it, and . . . none but the ignorant vulgar give it any credit') in 1708, a change is already perceptible: from enthusiastic (dangerous, radical) into vulgar (common, crude). By the early eighteenth century, no one could have seriously represented astrologers as a real threat to the state or civil order; yet it had become a mental habit of the educated elite to scorn them. Despite increasing distance from the original concern, the door to astrology remained as firmly shut as ever. When Erskine attacked the Company of Stationers' almanacs in Parliament in 1776, the target of his rhetoric was not their seditious radicalism, as it would have been a century earlier, but their tasteless and senseless vulgarity. Similarly, the *Tatler* and the *Gentleman's Magazine* carried a discussion of John Partridge in 1785–6.[32] On a point of biography, Bishop Percy of Dromore suggested that Partridge's true surname had been one assigned to him by the latter's old enemy, George Parker: namely, Hewson. Parker's intent, however – one which his readers at the time grasped perfectly well – had been to associate Partridge with the ferocious Protectorate justice, sectarian and regicide John Hewson (d.1662). The allusion was quite lost eighty years later, along with most of the associated feelings and ideas; but Partridge, although now a figure more of curiosity and ridicule than fear, was no nearer rehabilitation. In other words, what had begun as ideological had become a class mentality. (In the nineteenth century, it subtly altered again, acquiring an emphasis on 'ignorance' and 'superstition' – not as an offence to true religion, but to true, that is scientific, knowledge.) As a corollary, no great disputes or polemics were needed to keep astrology in its place; maintained by the subsequent hardening of the patrician– plebeian divide, the ban functioned more or less automatically and unconsciously.

Although more difficult to pin down, a slow change is also perceptible in the popular astrological mentality. Despite an appearance of timelessness, it too was unavoidably affected by the development of class-consciousness.[33] I have already characterized this as the shift from popular astrology to plebeian, in the course of which a well-thumbed copy of *Moore's Almanack* became the emblem of labouring people's attachment – increasingly static and repetitive, because defensive – to the rhythms and remedies of 'natural' time. Under attack both as astrology per se and by employers seeking to institute a different sense of time oriented to labour-discipline,[34] this mentality was chiefly characterized by resistance. As we have seen, popular astrology survived in eighteenth-century England much more successfully than did either judicial (horoscopic) astrology or the still more learned philosophical kind. It seems likely that the pressure from the top down itself generated a measure of resistance, whose passivity, far from being merely chthonic or inchoate inertia, was precisely its own best and most appropriate kind of defence.[35]

Significantly, the form of these most enduring astrological beliefs, in their very crudeness and simplicity, was far removed from the elite discourse of those attacking astrology; whereas judicial astrology, cast in a more sophisticated and learned idiom, suffered correspondingly greater attrition. To borrow a distinction between 'ideas' and 'idiom',[36] we may say that plebeian astrology was cast in an idiom – at once social, a way of life, and intellectual, a set of ideas – that was largely alien to that of elite culture. For that reason, it held no hope of taking on and actively contesting the latter's dominance; by the same token, however, and within certain limits – indicated by the Anglican and loyalist bounds within which *Moore's* kept itself – it was much better placed to resist encroachment. By comparison, the better-educated and socially higher judicial astrologers could hope to appropriate elite discourse in order to legitimate their astrology, and thus to argue their case on something approaching equal terms with their social superiors. But their vulnerability to influence (in this case, hostility), resulting from the proximity to that idiom, was correspondingly greater – a price reflected in the decline of judicial astrology.

By implication, mentalities and ideologies are best seen as inseparably linked ends of a continuum: the former more universal, habitual and assumed, the latter – because they are conflictual and ambitious – more particular, conscious and explicit. But the existence of pure types at either end may be doubted. Ideology is at the sharp end of changes in mentality, as we have seen. But mentality acts both to limit and undermine the former (when resistant), and to extend and cement it (when in sympathy).[37]

Apart from Vovelle, the historian who has recently worked hardest to recognize and theorize such complexities is Roger Chartier. Among his welcome emphases, which this work confirms, is the point that cultural type (such as elite or popular) is not some kind of essential identity, but stems from the use that a group makes of shared artefacts, practices or beliefs.[38] In this case, the shared beliefs comprise the fundamental astrological idea of cosmic relationship (effects, influence, sympathy) with earthly and human life – an idea that, formally speaking, was shared by people in all early modern social groups. The radical differences between those groups (without which this history would be reduced to triviality) lay in their different and often conflicting adaptations of that idea for their own purposes. Where its 'high' articulation implied a small, highly select, interpretive elite of Christian natural philosophers, the purpose of *Moore's* was to enable every person, with a minimum of assistance, to be his or her own interpreter. (As in other respects, the individualism of judicial astrology wavered in the middle, with the populism of a unique horoscope for everyone qualified by its interpretation, ideally, by an expert astrologer.)

Chartier's term for this process is 'appropriation'. Of course, it is true that, in the words of Carlo Ginzburg, dominant and subordinate cultures 'are matched in an unequal struggle, where the dice are loaded.'[39] The history of early modern astrology bears that out. As its neglected survival shows, however, there has been a tendency for historians to acquiesce in the self-interested verdict of the victors and their heirs by overestimating their success.[40] The idea of appropriation recognizes the limited, but real, freedom of people, including the objects of 'reform'. In particular, it recognizes their power as active consumers (that is, quasi-producers) of ideas creatively to adapt, subvert and redirect them. In fact, as we found with elite, judicial and popular astrologies, it is *only* through such strategies of interpretation that the formal ideas acquire their particular meaning and significance.[41]

Hegemony

This view converges, in a highly promising way, with another stream of recent socio-historical thought. The Italian Marxist Antonio Gramsci was intellectually motivated by some very similar concerns, especially, in his case, the need to transcend the crude reductionism of much Marxist analysis without sacrificing its commitment to social and material considerations. Gramsci realized that opinions with the currency of 'common sense' throughout societies (that is, mentalities), far from being merely neutral or random, are often in the interests of the ruling classes. He was also able to admit, however, that such views are usually adopted as a result not of force but persuasion – that is, even allowing for the effects of duplicity and privilege (and here we must reject the temptation ultimately to know better), voluntarily. In countries with serious social inequalities but where the State is not maintained by pure force of arms (such as most 'First World' countries today), such ideas – although belonging to the civil or cultural sphere – can be crucial in maintaining its political rule. In the importance of this 'moral and intellectual leadership', which Gramsci termed 'hegemony', he saw the possibility of alternative hegemonic constructions, led by the hitherto subordinate classes, which would be more socially just. His concept of hegemony therefore moves away from the view of a class or group imposing a dominant ideology (whether as edification or victimization) on a passive populace with no ideas or resources of its own; and the accompanying view of any resistance as either lumpen waywardness (if the dominant ideas are deemed a good thing) or doomed (if a bad). Hope and choice, in context, are thus readmitted to theory.[42]

Gramsci also observed (as we have, in the case of astrology) that every religion 'is in reality a multiplicity of distinct and often contradictory religions: there is one Catholicism for the peasants, one for the petit-bourgeois and town-workers, one for women, and one for intellectuals which is itself variegated and disconnected.'[43] What then unites them? Precisely the conviction that they are connected. To the extent that that belief is undermined – as natural philosophers sufficiently convinced themselves and others that their cometography was not astrology – the entity's 'essential' identity changes too.

These considerations are fundamental to the early modern history of astrology. As we have seen, its transformation from the first half of the seventeenth century, when it unevenly pervaded English culture and society, to a hundred years later, when it was confined largely to the lower classes and actively disappeared elsewhere, involved precisely the (partial) ascendency of a set of hegemonic ideas; just as its survival embodied resistance to those ideas. They were unmistakably hegemonic, not only because it was in the patrician interest for people generally to spurn radical and populist astrologers, but because the consensus, where it obtained, soon became a matter of 'common sense'. Relatively suddenly, astrology was for the great majority of educated persons a self-evidently vulgar superstition. The discovery soon afterwards by the professional middle classes that astrologers were a dangerous and uncouth breed was accomplished precisely by the power of gentry cultural hegemony. But plebeian astrology marked its limits.

Matters are not quite so simple as I might have suggested, however. To speak without qualification of 'the hegemonic voice of the gentry' disguises the fact that the power of the latter in the eighteenth century, although by far the dominant partner, was not unalloyed. Increasingly after the late seventeenth century, it was purchased at a price set by the historic alliance of the gentry with the middle classes. The latter had their own interests, which overlapped greatly with those of their social superiors but were not identical, and therefore provided potential leverage for criticizing the latter. Astrology was one subject which united them through a shared interest, but that did not rule out middle-class criticism of aristocratic corruption or inanity, such as that of Hogarth, or Swift and the Scriblerians. Towards the end of the eighteenth century, middle-class Enlightenment radicals, such as Bentham and Priestley, appeared on the scene. They undoubtedly shared the gentry's view of popular astrology as ignorant superstition – but extended it to Anglicanism as well. Another complication was that although different groups of reformers often worked in alliance, reform itself was constructed out of different and sometimes conflicting aims: those of employers to instil work-discipline, religious proselytisers to convert people, and moral–intellectual reformers to improve them.[44] That alliance was therefore strategic and imperfect, with a cross-class character, the essential beliefs of which 'were not the prerogative of any one class nor dictated [solely] by the interests of any class'.[45] Thirdly, some major reforms were undertaken by people of the same class background as the objects of their attention. Methodism in particular 'was at the same time of popular culture and opposed to it'.[46] That may seem a simple matter of hegemony (that is, the voluntary adoption of ideas implicitly supporting the upper classes) until one considers that those movements were often quite as offensive and alarming from the patrician point of view as were the objects of their reforms.

What all this suggests is that Gramsci's concept of hegemony, while representing a major improvement on the reductionist Marxism that preceded it, does not go far enough. Writing as political theorists, Ernesto Laclau and Chantal Mouffe have recently developed it further in the same direction. Briefly, they suggest that hegemonic ideas are not necessarily constituted around one of the 'fundamental classes' (and therefore that class is not intrinsically a more important consideration than gender, race, relationship with the environment and so on). By the same token, more than one hegemonic centre is possible. Hegemony is something unstable and unpredictable, and no particular kind is automatically privileged or guaranteed. Its dynamics are based on a 'logic of articulation and contingency', in which ideas derive their social meaning not from their origin, but from the way they are linked up with (that is, articulated) and embedded in other ideas, in the course of alliances based on various and shifting interests. There are limits to the ways and extent to which ideas can be articulated, but those limits are not inherent; they derive from the specific and contingent circumstances.[47] (As all this suggests, hegemonic struggle precedes and imperfectly produces hegemony just as class struggle does class, and for the same fundamental reason: because, like class, it is not a thing, but a relation.[48]) A hegemonic situation is therefore one where there are competing – not simply co-existing or complementary – interpretations of a shared set of ideas, which different groups are struggling to articulate differently for their own purposes. To the extent that a particular interpretation is successfully established as true, the result is 'common sense'. But that extent is never complete.[49]

It seems to me that this concept, together with that of appropriation, has a real contribution to make to the social history of ideas. Our understanding of the early modern history of astrology, for example, would be seriously impoverished without it. The alternatives of domination on the one hand and independence on the other are too Manichean; hegemony supplies the missing link.[50] We cannot expect to find pure types; there was naked coercion in the late seventeenth century, and it remained a constant possibility thereafter, mixed with lengthy periods of semi-independence in the eighteenth. But another, essential, aspect of this history is the way in which the idea of cosmic effects and influences was articulated so as to destroy the authority of astrologers in favour of a 'sound' interpretive elite. That campaign was hegemonic not only in its interestedness and its commonsensicality, but its cross-class character. As we saw with the development of Boyle's and Mead's cosmic physic, or the construction of comets as astronomical and eschatological (but not astrological) objects, it was a remarkable success. As we also saw, however, its success was incomplete; plebeian astrology survived, even flourished, none the less.

In sum, the human meaning of astrological ideas derived not from some transcendental essence, but from the way in which they were appropriated and articulated. The relationship between the different interpretations was potentially hegemonic just in so far as the 'truth' of one ruled out that of the others. But early modern astrology became the site of an actual hegemonic struggle only with the successful campaign within the patrician bloc, and the unsuccessful one in relation to the plebeian, when efforts were made to institute an elite interpretation more generally still. As we saw, those efforts were led first by theologian–natural philosophers, anxious to colonize astrology for natural philosophy, in the late seventeenth and early eighteenth centuries; then (overlapping but slightly later) by the literati, striving to establish periodical literature, as opposed to the almanac, as the model of legitimate intellectual authority.

One reason for their hegemonic failure *vis-à-vis* plebeian astrology was that, in the eighteenth century, little need was perceived (and therefore little effort made) to make any concessions in order to hegemonically convince – as distinct from simply dominate, threaten or contain – those beyond the patrician pale. That need, and the more complex concessions it entailed, did not occur until the following century. Where the campaign against astrology failed, then, it did so due both to a degree of genuine plebeian independence, or co-existence; and (in the face of direct attacks) to a resistance based on differing interests, on the part of people who quite rightly perceived no common cause with the ambitions or grievances of the critics. That resistance was necessarily defensive, and included some concessions: William Lilly's heirs distanced themselves from the people as he had not, and Moore's, unlike Interregnum almanacs, remained loyal. But it was not a rout, and plebeian culture continued to include a demotic–democratic potential that proved a seed-bed of later working-class self-discovery.[51] Patrician insularity generally, however, plus their success in eliminating astrology within their own ranks and confining it to 'no one worth speaking of' – cemented by the mental habits of a class mentality – meant that astrology was not a high priority for later reformers except indirectly (such as substituting a sense of work-time for natural time). The remainder of the eighteenth century was therefore

166

characterized overall by uneasy co-existence outside the patrician consensus.

That pattern was next broken in the early nineteenth century. Alarmed by a surge of middle-class (that is, directly competitive) interest in judicial astrology, scientific writers and publicists, and the leader-writers of the *Athenaeum* – direct heirs of astrology's critics a century earlier – published a fresh spate of bitter attacks. Presaged by the success of Sibly and Worsdale in the 1790s, this new interest was a result of the further fracturing of gentry hegemony itself in the turbulent 1820s and 1830s. It centred on the figures of 'Raphael' (Robert Cross Smith) (1795–1832) and 'Zadkiel' (Robert James Morrison) (1795–1874), whose busy practices and new almanacs, selling annually about 10,000 copies each, found an enduring niche in Victorian culture and society. Their attempts to go farther and breach the upper-class cultural stronghold failed; but later in the century Alan Leo (1860–1917) was able to build on this base, and he created an eclectic and commercialized astrology (under the aegis of the Theosophical Society) that is still with us today. In the meantime, working-class autodidacts were beginning to lay hold of Enlightenment ideology for their own purposes; and these pioneers invariably rejected astrology. They were untypical, and sales of *Moore's* slowly rose to a peak of over half a million in 1839; but decline set in thereafter, and towards the end of the nineteenth century popular astrology was at last disappearing from the countryside. (Whether or not it did so from the towns, it certainly reappeared, led by the *Daily Express*, in the 1930s).[52]

Thus the story which we have followed from 1642 to 1800 continues, and remains a challenge to our historical self-understanding. Looking back at the early modern period, it is sometimes difficult to avoid seeing astrologers as swept along by inexorable historical forces, a tide of change that left them only with the choice of whether 'to shout and thereby hasten the end, or to keep silent and gain thereby a slower death.'[53] We have seen the irresolution of that death, however, and the new (if changed) life that followed it. I have tried to show the contingency of choices and events – structured, certainly, but only by processes themselves contingent; and determining as well as determined. Despite some of the grander claims made in the name of their art, astrologers and their contemporaries alike had no certain foreknowledge of the future. In any case, no outcome was ever single or forever fixed, least of all in its human meaning. There was always, therefore (at least in principle), everything to play for; and in fact we found no passive pawns of history, but people active in the pursuit of their ends, making use of their historical resources as well as being shaped and directed by them.[54]

That much remains unchanged. We too face the interplay of past and present, simultaneous gain and loss, new threats and new possibilities. Far from undermining our abilities to understand and act, I believe that honouring 'the essential relativity of things human', in all its ambiguity and ambivalence, has powerfully positive effects. In this view, history is a place 'where no one owns the truth and where everyone has the right to be understood.'[55] Thus a persistent humility is required; for no part of the human past is beyond question, or unworthy of our attention.

Notes

Notes to chapter 1

1   Howe (1984) 5.
2   The Irish critic and philosopher Richard Kearney.
3   In a letter to Elias Ashmole (12 February 1666), Ash. MS 423, ff. 256–7.
4   Standard accounts are Thomas and Capp; see also Leventhal (1976), Webster (1982), MacDonald (1982), Monter (1983), Howe (1984), Hunter and Gregory (1987), papers in Curry (1987), and theses by Bowden (1974), Wright (1984), Curry (1986) and Geneva (1988). Cf. Obelkevitch (1979) 7, who points out that especially neglected 'is the question of the pace and process of change in popular religious life between 1650 and 1800: between the major phase of the Reformation and Counter-Reformation and the advent of the modern economic and political revolutions.'
5   Thompson (1968) 12. An attitude also encapsulated for our times by Douglas Adams, in *The Hitch-Hiker's Guide to the Galaxy* (1979), 89: 'Many men of course became extremely rich, but this was perfectly natural and nothing to be ashamed of because no one was really poor – at least no one worth speaking of.'
6   On history and theory, see Thompson (1978a), Abrams (1980, 1982), Burke (1980), Samuel (1981), Johnson et al. (1982), G. S. Jones (1983), Neale (1983) 271–307, Veyne (1984) and Vilar (1985).
7   Thomas 797; cf. Capp 277.
8   Power construed as formative and enabling, as well as negative and repressive; see e.g. Foucault, in Dreyfus and Rabinow (1982), but also (qualifying Foucault) Said (1984) 243–7 and Laclau and Mouffe (1985) 142.
9   Recent examples are Williams (1984), Cooter (1985) and Barrow (1986).
10 On anachronism and 'Whig history' – which he termed 'the ratification if

not the glorification of the present' – see Butterfield (1931), and the excellent discussions in Ashplant and Wilson (1988) and Wilson and Ashplant (1988). It can take many different forms, from conservative through liberal to Marxist, and religious to scientific; but in all cases, the assumed privilege of the writers, the resort to historical anachronism, and the assumed teleology of history are reprehensible. It may be objected that I have used anachronistic terminology myself, such as 'class' and 'ideology'. I have not done so, however, in order to put constructions on their actions which my historical subjects would have failed to recognize in their own terms, nor (above all) to praise or blame them.

11 Obelkevitch (1979) 5; see also Chartier (1984) 230, and Vincent (1989), who correctly defines superstition as 'a pejorative description of any belief with which the user disagrees.' Cf. Cooter (1984) 35: 'the task before the historian of so-called pseudoscience is not to make further privilege for positivist science, but to determine how and why some conceptions of reality acquire the mantle of objective scientific truth and enter the domain of common sense, while others come to be regarded as arrant nonsense.' (That includes attempts to show, for example, that phrenologists were 'on the right track', 'partly true', and so on – thus deflecting attention from the interests concealed in the 'truth' of modern science itself; also see Cooter (1976; 1985).) Examples of the kind of treatment being criticized here include Sarton (1952), Jones (1961), Shumaker (1972), Rowse (1974) and Cowling (1977) on astrology (but nothing as crass, fortunately, as Graubard (1985) on witchcraft); and, more generally, the introduction to, and paper therein by, Vickers (1984), on which see Curry (1985). In a modern context, see also Feyerabend (1978) 91–5. A recent example (admittedly in a semi-popular book) is that of Couttie (1988) 72, who writes that 'It all comes down to whether astrology actually works.' This is just what it does not all come down to, since any answer to such a simple-minded question (whether yea or nay) is not only problematic, but at best unhelpful for understanding astrology's human and historical meanings.

12 Lloyd (1979) 263–4. Cf. the 'strong thesis' in sociology of science, in Barnes (1974; 1977), and Barnes and Bloor (1982).

13 See Gramsci (1971) 420, and Laclau and Mouffe (1985) 14: 'Plurality is not the phenomenon to be explained, but the starting point of the analysis.'

14 Notwithstanding the idealism of Febvre and Ariès, and some interpretations of post-structuralism, including that evident in Darnton (1984; 1986).

15 See Laclau and Mouffe (1985).

16 Thomas 4–5, Wrightson (1982) 17–38 and (1986) – especially the discussion of class in early modern England on p. 198 – and Clarkson (1971) 36–7; for further discussion and references, also see chapter 7. A recent guide is Sharpe (1987).

17 Stone (1980) 174, Holmes (1982), Cressy (1980) 73–4.

18 Thomas 767–800.

19 Capp 14.

20 Metaphors of death abound, e.g. Graubard (1958), Thomas 418, 424, Capp 278, Tester (1987) 240, 242.

21 e.g. Garin (1983), Eade (1984), Clarke (1985), Lilly (1985), Tester (1987); the standard modern edition of Ptolemy is (1940).

22 Otherwise fine, but mistaken on this point (which leads him wrongly to apply horary rules to a nativity), is Eade (1984) 39, 95–7.

23 Neugebauer (1953), Thorndike (1955).

24 Cornelius (1983/85); Tester (1987), *s.v.* 'katarchai'.

25 Butler (1680) C3v.

26 North (1980).

27 North (1985).

28 A letter in the *Times Literary Supplement* (29 January–4 February 1988); and see Geneva (1988). One place to start, as she has suggested, is to ask the following question: just what was it that, in the opinion of so many contemporaries, William Lilly (or Henry Andrews, or Ebenezer Sibly) was so good at doing?

29 Cf. the useful points made in Eade (1984) 2.

30 See Curry (1985), Schaffer (1985).

31 Cf. Eade (1984) 86–8.

32 Ptolemy (1940) 13.

33 See Kuhn (1962), and his 1987 Sherman Memorial Lectures in London; also Hesse (1980), and the journal *Social Studies of Science* generally.

Notes to chapter 2

1 Ash. MS 423, f. 274, Warren (1651), Homes (1652) 57, Gataker (1651) 188.

2 Hill (1974) 154–5, (1985) 40.

3 Heywood (1904) 6, Nicolson (1939) 2.

4 Hill (1985) 37.

5 Hill (1985) 34, 51, Greg (1956) 13–14; Gresham College was also empowered in this respect.

6 Stahlman (1956), Wing (1945–51).

7 Preface (n.p.).

8 Lilly (1985), originally (1647), Culpeper (1652b), Gadbury (1659), Coley (1669), Blagrave (1672), Saunders (1677), Salmon (1681), Eland (1694).

9   Capp; also Plomer (1885), Bosonquet (1917, 1930), Blagden (1958, 1961).

10 Blagden (1958) Table 1, Capp 23, Thomas 348–9; cf. Bosanquet (1930) 365, Blagden (1958) 115–16; for Lilly, see J[ohn] A[llen] (1659) 15 and Plomer (1906) 35.

11 See e.g. Hill (1975), McGregor and Reay (1984).

12 See McKeon (1975), Geneva (1988).

13 Booker (1646), *DNB* II, 830, Capp 174, 298.

14 *DNB* V, 286–7, Capp 303; quotation from Barclay (1985) 237.

15 Rowse (1974), *DNB* VII, 438–41, *DNB* VII, 461, Capp 306–7, MacDonald (1982) 16, 48–54; cf. *DNB* XIV, 71–3.

16 *ESTC*.

17 September 1649; quoted in Poynter (1962) 159.

18 Culpeper (1654b) 64.

19 On Blagrave, *DNB* II, 619–20, Capp 297; on Saunders, *DNB* XVII, 817–18 (although mistaken on date of death), Capp 329.

20 Thomas 446–7.

21 Rowland (1651), Epistle (n.p.).

22 Lilly, *Merlinus Anglicus* (1655) A4v.

23 Boehme (1656) 583–4.

24 Bourne (1646) T1.

25 See Thomas 443–9.

26 Spittlehouse (1650) B3–4v, *BDBR* III, 194–5, Gadbury (1689) 105, Ash. MS 427, f. 59v, *BDBR* I, 185, Capp 187–8, *BDBR* I, 133–4.

27 Ash. MS 420, f. 267, Thomas 372, *BDBR* II, 277–9.

28 *BDBR* II, 186–7, Thomas 443–4, Ash. MS 387, ff. 86, 217, 385, *BDBR* III, 78, Clarkson (1660) 32.

29 Stokes (1652) 22.

30 *BDBR* III, 329–32, Winstanley (1941) 578.

31 Webster (1654) 51; and see *DNB* XX, 1036–7, Elmer (1986).

32 Ash. MS. 240, ff. 205–6v, 76–7; Ash. MS. 243, f. 173v; Ash. MS 240, ff. 117, 119. Lilly was also consulted by Major-General John Lambert, Lieutenant-Colonel Read, and Sir John Reynolds. Other examples include Roger Crab (Ash. MS 210, ff. 107v and Ash. MS 427, f. 51v), Colonel Thomas Morgan (Ash. MS 241, f. 30v), Nicholas Gretton (Ash. MS 423, f. 134v), and Adjutant Allen (Ash. MS 210, f. 134v).

33 *DNB* XX, 1313–15, Lilly (1974) 63, Capp 337.

34 Wood (1691–92) IV, cols. 5–9, quoted in Josten 357, n. 3.

35 See Curry (1988).

36 Curry and Cornelius in Lilly (1985), Lilly (1974), Parker (1975), *DNB* XI, 1137–41.

37 Lilly (1985) 439–42.

38 *DNB* XXI, 112, *BDBR* III, 316–18.
39 Lilly (1974) 57–8, 62–3, 43–5, 64, 67–71, (1658) A7v, (1650) A3v.
40 Thomas 362–82, 379, 380–1, Parker (1975) 179.
41 Ash. MS 423, ff. 168–9, 173, L.P. (1652) A3v, Evelyn (1879) III, 144; cf. Warren (1651) 16.
42 Notwithstanding Halbronn's (1987) interesting points concerning Lilly's sources.
43 Lilly (1985) B.
44 Cornelius, in Lilly (1985) 866.
45 Lilly (1974) 32–3, 88, and Curry, in Lilly (1985) 861.
46 Christianson (1968). Field (1984).
47 *DNB* XVII, 1283, *DSB* XI, 342–3, Capp 330–1, Taylor 235.
48 Jones (1961) 123–4.
49 Shakerley (1649), Preface, 3.
50 Shakerley to John Matteson, *Historical Manuscripts Commission. Report on Manuscripts in Various Collections* (1913) VIII, 61.
51 Wing and Leybourne (1649a), Shakerley (1649).
52 Ash. MS. 423, ff. 111–29; on Shakerley and astrology, see also Kelly (1977) 61.
53 *DNB* XIV, 446–7, Capp 339, Taylor 222–3; Flamsteed quoted in *DSB*, XIV, 446–7.
54 For balanced views, see Brackenridge (1979), Field (1984; 1987[a] and Rosen (1984).
55 Wing (1651; 1656; 1669), Wing and Leybourne (1649a).
56 Wing to John Booker, Ash. MS. 423, f. 99, Gadbury (1669).
57 *DNB* XIV 446–7.
58 Ash. MS. 423, ff. 235–7, Whiteside (1970).
59 Streete (1653) 'To the Reader' (n.p.); on Streete and astrology, see also Kelly (1977) 61.
60 Josten, Lilly (1974), Hunter (1983), and *DNB* I, 316–18.
61 Hunter (1983) 5.
62 Ashmole (1652) 445; cf. Warren (1651) for a very similiar definition; and see Hoppen (1976).
63 *DNB* IX, 768, Josten 363, n. 7.
64 Ashmole (1652) 443, Josten 468–9.
65 Hunter (1985) 5–6.
66 Ashmole (1650), frontispiece; cf. Lilly's portrait in his *Christian Astrology* (1985).
67 Ashmole (1652) 453.
68 Lilly (1985) B2; Culpeper (1652) title-page.
69 See Logie Barrow's excellent discussion of democratic and elitist epistemologies in his *Independent Spirits* (1986) 146–61. (I have not

emphasized Ashmole's sexism, which is clear enough in the quotations given; although it was undoubtedly typical in this period.)

70 Ash. MS. 436, Josten 173, 184, 188–9, 1145, 1296, 1347–8, 1350–1, 136, Hunter (1983) 40–1.
71 *DNB* IX, 768–9.
72 Heydon (1662a) and (1664) C3v.
73 *DNB* IX, 769.
74 Gadbury, . . . *A Diary* . . . *for* . . . *1694*, Av.
75 Josten, *s.v.* 'Society of Astrologers', Pepys (1970–83) I, 274, and Curry (1988).
76 Lilly, *Merlinus Anglicus* (1648) 31.
77 Besides Lilly and Ashmole, they include Wharton, Booker, Culpeper, Gadbury, Saunders, Salmon, Blagrave, Wing, Oughtred, Fiske, Ramesey, Atwell, Goad, Childrey, Edlin, Streete, Coley, Heydon, Jeffrey Le Neve, John Evans, Francis Bernard, John Butler, William Eland, Robert Turner, Nathanael Nye, Hardick Warren and John Rowley, Jr.
78 Englefield (1923) 143.
79 Josten 1490.
80 Ashmole (1927) 45.
81 *DNB* VII, 994, Thomas 448.
82 *DNB* XVI, 848–49.
83 *DNB* XIX, 188–89.
84 Carpenter (1657) 7, *DNB* III, 1073.
85 For an exhaustive list, see Norris Purslow's in Wellcome MS 4021.
86 Josten 1485, 1705.
87 *DNB* XIII, 1139, Taylor 233–4; Hunter (1982) *s.v.* 'Joseph Moxon'.
88 Just to speculate, it probably included – besides Gadbury, Salmon, Coley and others of old – Partridge, Moore, Parker, and possibly Richard Kirby, John Bishop, John Merrifield and John Holwell.
89 Josten 250.
90 On Dering, see Josten 241–2, 1678, 1712, Gadbury (1684) (n.p.), Partridge (1697a) and his *Merlinus Liberatus* (1697) and Ashmole (1927) 116.

Notes to chapter 3

1 Lilly (1974) 84–5, Hall (1654) 199.
2 Ramesey (1653) 225.
3 Evelyn (1879) 3, 144.
4 Hill (1980) 46; for a good discussion, see Hill (1985) 33–71.
5 *DNB* XI, 997–1007, 1000, Capp 48–9.

6 Muddiman (1923) 153–6.
7 *CSPDS* (1661–2) 23, 572, 589, 592, *BDBR* I, 119–20, Muddiman (1923) 153–61, Shapin and Schaffer (1985) 291–2.
8 *DNB* IX, 768–9, *CSPDS* (1663–4) 229, 230, 246–7.
9 *CSPDS* (1666–67) 541.
10 Hooke (1935) 77–9, Hill (1985) 52, Shapin and Schaffer (1985) 292–3.
11 Alexander (1968) 220, Hill (1980) 55, Holmes (1982) 4.
12 Lilly (1974) 101.
13 *DNB* XV, 272–5.
14 Parker (1666) 2, 72–3, *DNB* XV, 272–5.
15 Casaubon (1668) 141, *DNB* III, 1170–1.
16 More (1661) 134, 53–4, *DNB* IX, 509–10.
17 See Jacob (1974).
18 See e.g. Henry (1989).
19 Parker (1666) 90.
20 Allen (1973) 243, *DNB* III, 526–8, Butler (1663–78); see also Nelson (1976).
21 On Wilson and Congreve, see Eade (1984) 200–6, 211–15.
22 Epilogue to *Sr. Martin Mar-All*, acted 1667, quoted in McKeon (1975) 230; Eade (1984) 207–10, McKeon (1975) 261–2; Ash. MS. 243, f. 209, Dryden (1821) 18, 133.
23 Quoted in Heywood (1904) 77.
24 See ch. 2, n. 12.
25 No. 48, 26–30 April (1666).
26 Lilly (1974) 89–90, Ash. MS 553, ff. (8) and (11) and (13) on the plague and fire respectively; there is a tantalizing handwritten note in the margin of the last which reads 'forsan [perhaps] 1666 vel [or] 1667'. See also Thomas 4.
27 Edlin (1664) 42, 72, 118, 115, *DNB* VI, 389, Josten 104.
28 Nicolson (1965) 126, Lilly (1974) 101, 103–6.
29 Capp 50, *DNB* XI, 997–1007.
30 On Toland, see *DNB* XIX, 918–22, Pocock (1975) 403, Jacob (1976a).
31 Josten 1485, Gadbury (1669); cf. Sloane MS 2283, f. 13.
32 Blagrave (1672) Epistle (n.p.).
33 Butler (1680) (n.p.).
34 Saunders (1677) 172.
35 *DNB* XIII, 796, Capp 320; but Howe (1981) 200 gives Moore's death as about 1724, citing Henry Andrews.
36 Wright (1979) 94.
37 *DNB* XVIII, 817–18, Capp 329.
38 Buck (1977) 67, Shapin and Schaffer (1985) 283.
39 Corrigan and Sayer (1985) 94, Shapin and Schaffer (1985) 304. Cf. Hill

(1965), Jacob (1974; 1977b; 1978; 1980), and J. R. and M. C. Jacob (1980). On the phrase 'authorized prophets', see ch. 6, n. 19.

40 The most thorough treatment of this subject is Bowden (1974); I found this very useful, although my own discussion (in some respects more detailed in Curry (1986)) is based on primary sources. I should also point out that Capp (p. 183) does mention a Whig–Tory split among English astrologers over the issue of Placidus's house system; however, he fails to draw out its extent or significance. 'Science' is a convenient but anachronistic term here, which I have occasionally made stand in for 'natural philosophy'.

41 Gadbury, ... A Diary . . . for . . . 1678, 3–4.

42 e.g. Goad to Gadbury (August 1684), Ash. MS 368.

43 Buck (1977) 67–84, 67, 80.

44 Debus (1970) 30, DNB XX, 793–7 and XI, 264–7, Hoppen (1976).

45 Sprat (1667) 364–5, DNB XVIII, 827–32, cf. Jacob (1980) 25.

46 Childrey (1653) 2.

47 All from Bacon's De Augmentis Scientiarum (1623), Bk 3, ch. 4, published in translation in Bacon (1858) IV, 349–55; Bacon, Historia Ventorum (1622) in (1857) II, 19–78; Jenks (1983).

48 See Bowden (1974) 108–29, Brackenridge (1979), Simon (1979), Apt (1982), Field (1984).

49 Letter (9 March 1666) reproduced in Boyle (1744) I, lxxix. Boyle is almost certainly referring to John Heydon; see Jacob (1974; 1977a, b).

50 Boyle (1744) IV, 85–6, IV, 98, Boyle Papers, VIII, 204–5, reproduced in Curry (1986) 262.

51 Boyle (1744) 638–9, 642.

52 Petty (1647) 11–12; on Worsley and astrology, see Curry (1986) 125–7, 258–61; and Webster (1975).

53 Wren (1750) 203.

54 DNB IV, 250–1, Capp 301, Taylor 226–7; Oldenburg (1968) VI, 108–9; V, 454, Phil. Trans. (10 October 1670) V, 2061–74.

55 Childrey (1652) 3, 9, 3, 8, 13.

56 Childrey (1653) 4.

57 Childrey (1660) B5v–6v.

58 'Fitzsmith' may be a pseudonym for Childrey; see Capp 306, Nicolson (1939) 4–5.

59 e.g. Gadbury, 'A Brief Account of the Copernican Astrology', appended to A Diary . . . for . . . 1695, Capp 301; Phil. Trans. 2 (1667) 445; Oldenburg (1968) VI, 108, 110.

60 Wood (1820) IV, 267, Bodl. Vet. A3.e.173 and Pamphlet c.118.(17.), DNB XXII, 18–19; cf. Bowden 176–7.

61 Bowden 63–77, 109–28.

62 As Bowden 178 points out, in 1673 John Beale, a Fellow of Royal Society

who was indeed (as Gadbury put it) 'Astrologically as well as physically inclined', advocated compiling a 'Kalender' of the weather, epidemics, etc., and analysing it, after the manner of Graunt, to see 'how far the Positions of the Planets, or other symptoms or comcomitants, are Indicative of the Weather'. (*Phil. Trans.* (13 January 1672–73) VII, 5138–43.) She further suggests that Goad may well have followed this letter up and adopted some of Graunt's statistical methodology in the latter's pioneering *Natural and Political Observations made upon the Bills of Mortality* (1662). However, Goad began keeping his weather diary two decades earlier.

63 Gunther (1923–45) XII, 305. Locke's records, 1666–83, were an exception; see Bowden 179. Sloane MS 1731, f. 32; Ash. MS 368.

64 Sloane MS 1731, f. 32.

65 Ash. MS 368, f. 303b.

66 Goad (1686) 15.

67 Goad (1686) 57, 39–40, 37.

68 Goad (1686) 59.

69 Hacking (1975) 28, 39, cf. Kargon (1963), Leeuwen (1963), Sheynin (1974).

70 Hooke (1935) 204.

71 Birch (1756–7) III, 454–5.

72 Royal Society MS 243, no. 35; see Hunter (1987).

73 Gunther (1923–45) XII, 203, Plot in *Phil. Trans.* (1684) XV, 930–1.

74 *Acta Eruditorum* (January 1688) 22–4, my translation; Goad (1686) 490–1.

75 Gadbury (1660), (1665b), (1674) and (1689) respectively.

76 Gadbury, . . . *A Diary* . . . *for* . . . *1689* (n.p.), . . . *1695*, A6, and . . . *1662*, 186, Tanner MS 22, f. 126, Josten, s.v. 'Gadbury'; *DNB* VII, 785–6, Capp 308.

77 Aubrey (1898) I, 35, I, 9, Aubrey MS 23, Aubrey (1898); see Hunter (1975).

78 Oldenburg (1968) II, 532, Hooke (1935) 248, 'Espinasse (1956) 119, Hooke (1935) 86, 102, 204, 248, 386, 67, 156, 176, Gunther (1935) X, 76.

79 Cohen and Ross (1985) 2–3.

80 Ronan (1966) 47.

81 Gadbury, . . . *A Diary* . . . *for* . . . *1670*, C3–4, and . . . *1676*, C7–8.

82 Gadbury, . . . *A Diary* . . . *for* . . . *1664*, Av, and . . . *1671*, A5.

83 Gadbury (1689) and . . . *A Diary* . . . *for* . . . *1703*, A1–3.

84 *DNB* XV, 233–4, Capp 249, 322–3.

85 Parker, *Mercurius Anglicanus* (1690), Hunter (1987) 265.

86 Hunt (1696) Preface and title-page, Taylor 267; Hunt *fl.*1673–98; Godson (1697) 1, 8, 40.

87 Younge (1699) A3–4.

88 Mayhew (1961) 37, *DNB* XV, 428–30, Capp 323, Taylor 209.
89 Mayhew (1961) 37, Partridge (1679) 2.
90 Partridge (1716).
91 Partridge (1693) title-page, 7, viii.
92 Partridge (1697a) A2, 3, 93–4, 62v, Bv, title-page, Bv, B2r.
93 Webb (1878) 560, Hoppen (1970) 16–17, 23, Capp 337, Evans (1976) 20–42.
94 Evans (1976) 30.
95 On the hyleg, see Eade (1984) 91–5; anareta is any planet in unfortunate aspect to the hyleg.
96 On Coats, see Evans (1976) 49–55; quotation from p. 52.
97 Thorndike (1923–58) VIII, 302; see also Placidus (1983), Introduction and Bibliography.
98 Placidus (1983) 47.
99 Placidus (1983) 'Primum Mobile' (n.p.); In the opinion of North (1985), Placidus's system of house division originated with Magini (Maginus), on whom see Clarke's excellent study (1985). (The same system is used today by many astrologers, having been popularized by the early Victorian astrologer Raphael.) Placidus (1983) 14.
100 Partridge (1697b) 28, Ptolemy (1701) 93, Parker, Merlinus Anglicanus (1697) C8r.
101 Gibson (1711) 2, 8, Capp 309.
102 But see below, ch. 5, on John Worsdale. Note also the reappearance in recent years of an ambitious programme, led by M. Gauquelin, to reform astrology statistically and scientifically; see Gauquelin (1984) and recently Seymour (1988). *Plus ça change?*
103 Pocock (1957) 47 and passim.
104 Jacob (1983).
105 Partridge (1693) 38.
106 Hunter and Gregory (1988).
107 See *DSB* IX, 527–8.
108 *Athenian Mercury* 1683, issue 6, cited in Hunter and Gregory (1988) 49–50.
109 Holwell (1682), Merrifield (1684).
110 Coley, *An Ephemeris . . . for . . . 1699*, 93v, *DNB* IV, 784–5 (but note that the date of death is in error), Capp 302, Taylor 241.
111 Wood MS F. 51, f. 6v.
112 The principal secondary sources are *DNB* XV, 428–30, Eddy (1932), Muddiman (1923) 248–55; the basic works are collected in Swift and Pope (1727) I.
113 Brown (1760) I, 146, 150, and II, 227, Eddy (1932) 34–5.
114 [Swift] (1708) 4.
115 [Swift] (1709) 5–6.

Notes to Chapter 4

1 The only major secondary sources on eighteenth-century English astrology to date are Capp 238–69 (which is very good), Howe (1984) 21–8, and Harrison (1979) 39–54.

2 Thrale (1942) II, 786.

3 Wharton (1648) A2; Coley, *Merlinus Anglicus Jr.* (1679) C7–8 and *The Family Almanack* (1752) 1; Season, *Speculum Anni . . . 1761* 2–3 and . . . *1762* 3.

4 Burke (1978) 281 and (1981).

5 Chartier (1982) 30.

6 On earlier English popular astrology, see Wedel (1920), Tomkis (1944), Brasewell (1978), Carey (1984), and Capp; on French, see Saintyves (1937).

7 Horn (1980) 13–14, Porter (1982) 25.

8 See Obelkevitch (1976, 1979); Barry (1986a).

9 Stone (1980).

10 Quoted in Dyer (1878)

11 Brand (1776) 380.

12 Thompson (1945) 12.

13 Harley (1885) 218, Brand (1905) II, 418–19.

14 Tusser (1931) 56, 96, Dyer (1876) 42.

15 Dyer (1876) 43.

16 Denam (1850) 11, Harley (1885) 216.

17 Henderson (1866) 86.

18 Latham (1868) 8, 11, 45.

19 Hone (1832) 254, Latham (1868) 30, Dyer (1876) 43.

20 Dyer (1876) 42.

21 Harley (1885) 202.

22 Brand (1905) II 420, Harley (1885) 195, 221, 185.

23 Dyer (1876) 38, Notes and Queries (1 August 1874) 84, Harley (1885) 186–8.

24 *Aristotle's Book of Problems* (?1710) title-page; *Aristotle's Masterpiece* (n.d.) title-page and 14; see Porter (1985b).

25 *ESTC*; Poynter (1972).

26 Salmon, London Almanack (1704), quoted in Bosanquet (1930) 391.

27 Younge (1699) Epistle Dedicatory (n.p.).

28 Blagden (1961), Cressy (1980) 177 and graph 8.1, Belanger (1982), Porter (1982) 381, Howe (1984) 21–22. On almanac sales also see Curry (1987) 266–7. On newspaper ciculations see Sutherland (1935), Snyder (1968), and Harris (1987). On *Poor Robin* see Capp 123–6 and *passim*,

and for another example of plebeian scepticism about astrology, Hobbs (1981) 45, 64–5 (drawn to my attention by Roy Porter).

29 Advertisement in Coley, *Merlinus Anglicanus Jr. for . . . 1708.*

30 Francis Moore's date of death is disputed; according to Henry Andrews (cited in Howe (1981) 200), Moore died in about 1724.

31 Blagden (1961) 21, 41–2, Howe (1981) 200–3, 207–9, Plumb (1982) 269, Porter (1985a).

32 Valenze (1978).

33 Quoted in Neuberg (1971) 94; Clare (1827) 2.

34 Knight (1864) I, 151.

35 [Man] (1810) 118–20.

36 Dawson (1882–3) 197–202.

37 Fletcher (1910) 102–9, Bamford (1841) 130–1, Harland and Wilkinson (1867) 123–5.

38 Southey (1807) II, 342–3.

39 Howe (1984) 23–4, Leventhal (1976) 56, *Morning Herald (October* 1786), cited in Howe (1984) 23, The Astrologer's Magazine (1793) 20.

40 Wellcome MS 4021 (with thanks to Michael Hunter for pointing this out to me).

41 Walton (1733) 12–17.

42 Charles (1716) 6; thanks to Keith Thomas for helping to locate this document.

43 *DNB* IX, 1250–2, 1251.

44 *The True Characters* (1708) 61.

45 *Tatler* (1709) No. 126, *Spectator* (12 October 1712) No. 505.

46 Bayle (1708), Labrousse (1983).

47 Harris (1704) I, 12r, Chambers (1728) II 388, *Encyclopaedia Britannica* (1771) I, 433; see ch. 6.

48 *Gentleman's Magazine* (February 1740) X, 76, and (September 1732) II, 974.

49 Boyse (?1760) title-page.

50 Johnson (1755), see 'Astrology' (I, n.p.) and 'Superstition' (II, n.p.), [Man] (1810) 22.

51 See Capp 434 (n. 12) and 438 (nn. 122–3), Blagden (1961).

52 Beaumont (1803) iii.

53 Harland and Wilkinson (1867) 121–2; cf. Peacock (1856) 415, Henderson (1866) xvii.

54 Payne (1979) 20; for a fuller discussion and references see ch. 7.

55 On popular religion, see Plongeron (1976), Duboscq et al. (1980), Ginzburg (1980), Hilaire (1981), Monter (1983). On the withdrawal of the church, see Thompson (1974), Hill (1980) 75–7, Porter (1982) 191.

56 Berman (1975); for example, Sir Hans Sloane, PRS 1727–41.

57 [Swift] (1708b) 3.

58 Cited in Malcolmson (1973) 166.

59 Hazlitt (1931) 91; Clare quoted in Paulin (1986).

60 Eagleton (1984) 12.

61 Eagleton (1984) 24.

62 See Olson (1983), Eagleton (1984) 19.

63 A phrase used by Thompson (1978) 162–3 and apparently endorsed in its essentials by Clark (1985) 88, 44; see Eagleton (1984) and ch. 7.

64 See Jarrett (1974) 183, Golby and Purdue (1984) 58.

65 See Malcolmson (1973) 102–7, Golby and Purdue (1984) 53.

66 See Thompson (1967), Vincent (1989).

67 Bamford (1841), Lovett (1876).

68 Sydney Smith, quoted in Golby and Purdue (1984) 52; see forthcoming work on the Proclamation Society by Johanna Innes.

69 Thompson (1974) 163.

70 On the remarkable longevity of an oral tradition among children, see Opie (1959), discussed more recently in Larkin (1983) 111–16. My wife and her friends, growing up in mid-Sussex in the early 1960s, were well acquainted with the idea of 'boggarts', or mischievous gnomes, part of popular magical culture (and by that name) since at least the seventeenth century.

71 This idea is based on Medick (1982).

72 Blagrave (1672) Epistle (n.p.).

73 Rule (1986) 226; see Obelkevitch (1976) 312 and Medick (1982).

74 Obelkevitch (1979) 5.

75 See Howe (1981), McKendrick et al. (1982), Porter (1982).

76 See ch. 7.

77 For example, *Vox Stellarum* 1788, 1789, 1792, 1793, 1794.

78 *Gadbury* (1736) C8v [Ephemeris].

79 Harrison (1979) 79.

80 Garrett (1975) 147.

81 Boehme (1764–81; 1770), Reid (1800), Hirst (1964) ch. 8, Garrett (1975; 1984), Harrison (1979), Schwartz (1980), and for a highly effective fictional portrayal of the millenarian milieu in eighteenth-century England, Fowles (1985).

82 Darnton (1968), Cooter (1985), Porter (1985a).

83 Harrison (1979) 221.

84 Thomas Foley to Charles Taylor (16 March 1821), BL Add. MS 47, 795, f. 94, Tozer [1812]; see Capp 252. I see no reason to avoid the word 'radical', J. C. D. Clark's 'ironical' but apparently indispensable use of it notwithstanding.

85 Thompson (1974) 162–4. For the moment, I shall defer a deeper discussion of hegemony until the concluding chapter.

Notes to chapter 5

1　See ch. 2.
2　Irish (1701) 53, 8.
3　Gadbury, *A Diary . . . for . . . 1703* A1–3, and . . . *1665* C7, Season, *Speculum Anni . . . 1774* C3.
4　*The Surprizing Monument* (?1715).
5　Beetenson (1722), Ball (1723), Beckworth (1733).
6　Ptolemy (1701; 1786), Phillips (1785), *Eland's Tutor* (1704), Ball (1723), Penseyre (1726), Heydon Jr (1785), Mensforth (1785). In England, up to 1695, about 100,000 titles were printed; the current *ESTC* stands at 200,000, and is expected perhaps to double (Robin Alston, editor of the *ESTC*, personal communication).
7　Howe (1981) 199; this was the number printed, not the number sold; however, we can assume that approximately the number must have been sold, since the same number was printed the following year.
8　Howe (1981) 199.
9　Saunder, Apollo Anglicanus (1715) C4, and cf. (1713) A5–B8, C3–5; Capp 246.
10　Lilly was born in Diseworth, Leicestershire; Edlin in All Saints Parish, Stamford, Lincolnshire (Wellcome MS 4729, title-page).
11　See Howe (1984) 26 and Capp 242; the term 'dynasty' is Capp's, too accurate to avoid repeating.
12　For brief biographies, see Capp 311, 331, 336, 338–40.
13　e.g. *Moore's Vox Stellarum* (1789) 43, and (1796) 40.
14　Examples of the latter: Thomas Allen (1542–1632), Nicholas Mercator (1619–87), John Napier (1550–1617) and William Oughtred (1574–1660); see Curry (1986) 81. Note also the existence of a tradition of excellence in mathematics in Lincolnshire, which supplied Isaac Newton with a solid education in arithmetic and geometry in the 1650s: Whiteside (1982).
15　*DNB* XIV, 127–9, Hoskin (1985), Piggott (1985).
16　Stukeley (1882–3) I, 104 and II, 304, BL MS 50148, 22v and 32r.
17　Wellcome MS 54729 (with thanks to Michael Hunter for pointing this out to me); axioms and diagram are in Stukeley (1882–3) I, 90–3.
18　*DNB* I, 406, Capp 294, 264–7, Howe (1981) 198, *Notes and Queries* (1851) IV, 74, 162, Kingston (1906) 212–22; obituary in the *Gentleman's Magazine* (February 1820); and an unpublished paper by Marc Paul Jones, to whom thanks for some of these references. The bulk of Andrews's astrological MSS appears to be lost, which is a great pity. He also edited *Remarkable News from the Stars* from 1769.

19 Capp 330, 259–62, Season, *Speculum Anni* (1761) 2–3, (1769) B8–C, (1762) A6, (1769) 3, (1771) 4; and, e.g. (1762) 7, (1772) C6v, (1774) C6; *Speculum Anni* (1761) A7r, A8r, (1777) C4v, (1772) C4r, (1733) A2v; *Speculum Anni* (1774) C2r–3r.

20 Hearne (1885–1921) VI, 97–98.

21 Gough (1981) 249.

22 Barry (1986) 151.

23 Woodforde (1931) I, 336, V, 1801, and see Hultin (1975) 362–3.

24 Thrale (1942) II, 851–2.

25 Stone (1980), Holmes (1982).

26 Barry (1986) 174 and *passim*; this makes more sense to me than Perkin's (1972) suggestion of an entrepreneurial–professional split within the middle classes.

27 Inkster (1977; 1983), Porter (1980) and (1982) 21, Russell (1983) 107 and *passim*.

28 Holmes (1985) 186; see also Darby (1983), Thirsk (1984) and especially Obelkevitch (1976).

29 Bowles (1977), Rousseau (1982), Secord (1985).

30 See ch. 6

31 *Encyclopaedia Britannica* (1771) 444.

32 Letter in the *Gentleman's Magazine* (1782) 411–12; see Hobhouse (1948).

33 Wilde (1980), Stewart (1981).

34 Brothers (1801) 82, 137, Prescot (1822) (mentioned in *DNB* II, 1350–3).

35 *Conjurer's Magazine* (March 1792) 340.

36 Obelkevitch (1976) 288–9, Howe (1984) 26–8.

37 Worsdale (1805) vi, x.

38 Worsdale (1819) 44.

39 Penny (1915) 8–9, 23–4; cf. Obelkevitch (1976) 288–9.

40 Worsdale (1819) 47.

41 Worsdale (1828) xv, v; Thomas White's *Celestial Intelligencer* and James Wilson's *Dictionary of Astrology* are nineteenth-century publications which are not discussed here.

42 *Conjurer's Magazine* (August 1793) 152, and cf. (March 1792).

43 *DNB* XVIII, 185, Ward (1958), Timson (1964–5), Debus (1982).

44 Debus (1982) 261 and *passim*; see also the BL catalogue.

45 Sibly (1784–8) III, 1050–1.

46 Sibly (1784–8) I, 20.

47 Adorno's phrase: (1974) 87.

48 Related briefly but very well by Howe (1984) 28–67. The subject was most recently given a boost by Ronald and Nancy Reagan.

Notes to chapter 6

1 See Introduction, n. 10; for an excellent counter-balance, see Webster (1982a).
2 On Newton, see Rattansi and McGuire (1966), Rattansi (1972; 1973), Webster (1982a), and Schaffer (1987a); a recent (if brief) historiographical discussion is Curry (1985).
3 'Espinasse (1956) 112–13.
4 Tong (1680); Tillotson to Nelson (5 January 1681), BL Add. MS 4236, f. 225.
5 [Edwards] [1684] 164.
6 Now published with a thorough and very useful commentary in Hunter (1987); quotations from pp. 288, 291.
7 Baily (1835) 34, and Newell (1984) 14 (with thanks to Annabella Kitson for the latter reference).
8 Flamsteed to Towneley (11 May 1677), RS MS 243, no. 26.
9 Cowling (1977) 2–3 (although otherwise an example of how *not* to approach the subject), Westfall (1980) 88, 98.
10 Schaffer (1987a), to which my discussion here is heavily indebted; quotations are from pp. 242–3.
11 See Dobbs (1975), Westfall (1984).
12 Keynes MS 19, f. 1r; quoted in Dobbs (1975) 154.
13 Thorndike (1955).
14 Hutchison (1982); and see Field (1987) 1, and Clark's excellent (1984).
15 Gregory (1702).
16 Christ Church College (Oxford) MS 113, ff. 47–50. (I am grateful to David Blow for drawing to my attention this passage of Gregory's, initially as mentioned in Eagles (1977) 105.)
17 See Gascoigne (1984) and Stewart (1986).
18 [Hammond] (1721) 7. (I am indebted to Angus Clarke for drawing this work to my attention.)
19 Lambert (1976) 63, as mentioned and discussed in Schaffer (1987a).
20 Ray (1713) 296–7.
21 Clarke (1730) VI, 152–3.
22 Whiston (1717) 23; see Force (1985).
23 See ch. 5, n. 4; Mackay (1841) 170.
24 Whiston (1696) 373.
25 Bowles (1977), Rousseau (1982), Secord (1985).
26 Derham (1715) 170–1.
27 Harris (1704) I, I2r, xxx4r.
28 Chambers (1728) II, 388; I, 162–3.

29 Diderot and D'Alembert (1751–65) I, 780–3, Collison (1964).
30 *Encyclopaedia Britannica* (1771) I, 433.
31 *Encyclopaedia Britannica* (1771) I, 444.
32 Ferguson (1754) 27.
33 Turner (1783) 20.
34 Hill (1754) *s.v.* 'Judicial Astrology'; Long (1742/64) 563.
35 *Encyclopaedia Britannica* (1771) III, 98.
36 *DNB* XIII, 181–6; Bowden (1974) 210–12.
37 Patte (1956).
38 Mead (1704) 20, 22–5, 27, 87; cf. Brookes (1771) I, 278, on the Moon and epilepsy.
39 Davy (1799) 145 (with thanks to Simon Schaffer for pointing out this passage).

Notes to chapter 7

1 Burke (1978); see also Kaplan (1984), Obelkevitch (1979). The phrase 'the reform of prophecy' is taken from Burke (1978) 274.
2 Rabb (1975), Hirst (1982).
3 See Capp 270–4, Halbronn (1987) 211–16.
4 e.g. Plumb (1963) 85, Thompson (1974) 403, Golby and Purdue (1984), Clark (1985) 68, 322.
5 Thompson (1974) 393, confirmed and developed by the careful work of Wrightson (1982) (see p. 227), also Payne (1979) and Obelkevitch (1979).
6 Thompson (1974) 395, (1978) 142, 151, Clark (1985) 88.
7 Clark (1985); see also Clark (1986).
8 See Brewer (1976), Colley (1982).
9 See Holmes (1982).
10 Brewer (1976) 267–8, Stafford (1987) 25–6.
11 Plumb (1963) 153; quotation from Porter (1982) 85; and see Thompson (1968).
12 Fletcher and Stevenson (1985) 12; and see Payne (1979).
13 Thompson (1978) 156, *against Thomas*: truer than he realized, since he mentions astrology as possible counter-evidence! (I have found no historical evidence whatsoever for any quasi-transcendent cognitive structures (mental, linguistic, or logical) that explain anything about the history of astrology.)
14 Thompson (1978).
15 e.g. Croix (1981; 1984).
16 As ably argued, in the case of 'horizontal' or community ties, by Wrightson (1986) and Rule (1986), and in the other cases by Laclau and Mouffe (1985).

17   Even one purporting to be a defence of Marx's theory of history, but wherein one may look long and hard for any of the latter: Cohen (1978).

18   See the critical literature on Althusser, especially that generated by Thompson (1978a), e.g. Johnson et al. (1982); see also Gramsci (1971) 407, Veyne (1978) 239–40.

19   Veyne (1978) 228.

20   Thompson (1977) 264. Out of an enormous literature, see also (on class in history generally) McLennan (1981), Neale (1981; and 1983), Abrams (1982), G. S. Jones (1983), Croix (1984), Hirst (1985a), Vilar (1985); (in early modern England) Thompson (1968; 1974), Payne (1979), Underdown (1980), Wrightson (1981; 1986), Medick (1982), Holmes (1984), Ingram (1984), Clark (1985).

21   Serious questions attach to Marxist perspectives on democracy and civil rights, which in my view still find better answers in modern political liberalism. However, this point fails to obviate the importance of class in history. *A fortiori*, any programme to overcome class divisions and their terrible effects cannot begin by ignoring them.

22   Especially the early *Annales* historians Lucien Febvre and Marc Bloch, together with Robert Mandrou, Jean Delumeau, and Robert Muchembled, and more recently Phillipe Ariès, E. Le Roy Ladurie and Jacques Le Goff.

23   For discussions, see Burke (1978a; 1987), Hutton (1981), Burghière (1982), Chartier (1982, 1988), Clark (1983), Gismondi (1985), Le Goff (1985).

24   Chartier (1982) 31.

25   Le Goff (1985) 168.

26   R. S. Jones (1983). (With thanks to Graham Douglas for pointing it out. NB: Jupiter, not Mercury, is the planet associated with the liver.) On planets and metals see Kollerstrom (1984), and as a mentality Burke (1986) 441. For a recent description of an astrological mentality, see Geneva (1987). Ives (1980) points out that almanacs 'taught the intimate relations between man and the universe, microcosm and macrocosm. Scientific advance might discredit the notion but people clung to it because it appeared to coincide with their own experience. Now, two-and-a-half centuries later, with the wreckage of our scientific hubris around us, we too are discovering "one world" and ecology.' The problem with seeing contemporary astrology as a democratic and/or green mentality, however, is its partiality; like ecology and democracy, what astrology 'is' depends on how it is articulated. (See below, in this chapter.)

27   Vovelle (1985) 324.

28   *Pace* the 'totalisante (on n'ose dire totalitaire)' and mysterious (not to say mystifying) globalism of Le Febvre and (especially) Ariès, who 'fait mouvoir sur coussin d'air l'évolution des attitudes . . . en fonction du dynamism propre d'un "inconscient collectif" non autrement défini' (Vovelle

(1985) 215, 234). This attitude has a genteel English cousin, which has been unkindly but aptly called 'antiquarian empiricism', in which any attempt at theoretical explanation is viewed as a suspicious alien import, changes are buried in 'thick description', and conclusions avoided. This is not necessarily harmful, except when the material demands them, as it does here. This danger has recently taken a new form in American academia, consisting of a tendency to indulge in arcane cognitive or symbolic meditations, usually para-linguistic and transcendental, with all political and social factors bracketed; for an example in historical writing (his otherwise excellent work notwithstanding), see Darnton (1985; 1986), for an incisive general critique, Said (1984), and more specifically Chartier (1988) 95–111.

29 *Pace* (for example) Georges Duby (1985) 151, 158 – although in other respects a valuable discussion – on 'cultural productions bearing only partial resemblance to material reality', and ideologies as 'the interpretations of a real situation'. As if in either case the former could ever be just copies, or the latter 'reality' could be discussed in any unproblematic way, and the two compared; as if we could ever know, let alone agree, whether the match was perfect. Along with variants of Marxist theory which contrast ideology – seen as 'false consciousness', or distortions of 'reality' – with (politically correct, proletarian, etc.) science, this is simply a Marxist version of scientific positivism, with equally unacceptable implications: an historiographical Hobson's choice.

30 Chartier (1982) 41, after Bourdieu (1979).

31 Baker (1982) 203; and see Williams (1976), *s.v.* 'Ideology', and Larrain (1979). To go any further is to invite the problems of moral and epistemological privilege just discussed; cf. ch. 2, n. 10.

32 Mayhew (1961) 38–42.

33 Obelkevitch (1976) 312; and see Thompson (1968).

34 Thompson (1967), Vincent (1989).

35 Cf. Obelkevitch (1979) 5.

36 Suggested by Proust (1981) I, 471.

37 See Vovelle (1985).

38 Chartier (1984) 235; also see Veyne (1971).

39 Ginzburg (1980) 155.

40 Chartier (1984) 235.

41 See Chartier (1982; 1984; 1988). He suggests the term 'appropriation' in preference to 'acculturation', from Muchembled (1978); the same point is made (in favour of 'negotiation') by Burke (1982) and Wirth (1984); cf. Gray (1976).

42 Gramsci (1971) 12 and passim. This is a large secondary literature on Gramsci's concept of hegemony, including e.g. Bates (1975), Joll (1977),

Mouffe (1979), Simon (1982), and Bocock (1986). Its importance in eighteenth-century England seems to be something on which Thompson (1975; 1978) and Clark (1985) are agreed, although both in a somewhat untheorized way which my account is intended to sharpen. For hegemony as a resolution of the structuralism-culturalism debate, on which see Johnson (1982), see the excellent introduction to Bennett et al. (1986) (with thanks to Steven Nugent).

43 Gramsci (1971) 420. On the same point respecting identity over time – that is, 'traditions' – see Hobsbawm and Ranger (1983), the brief but excellent discussion in Samuel (1984), and the introduction to Curry (1987).

44 Pointed out by Burke (1987) 212.

45 Golby and Purdue (1984) 86.

46 Golby and Purdue (1984) 58.

47 Laclau and Mouffe (1985) 137–8, 85, 168 and *passim*; see also Laclau and Mouffe (1987). I am grateful to Ernesto Laclau for some extremely helpful discussions of these matters. (It was fascinating recently to hear that scourge of post-Marxist heresy, Ellen Meiksins Wood, declare not only that 'post-Marxism' had not produced any good historical work, but that it *could* not.

48 cf. Thompson (1968) 8–11; and Gray (1976) 67 on 'the dynamic and problematic nature of the relationships' of hegemony, which 'applies more to a mode of organizing beliefs and values than to any particular set'.

49 cf. Williams (1977) 252: 'however dominant a social system may be, the very meaning of its domination involves a limitation or selection of the activities it covers, so that by definition it cannot exhaust all social experience, which therefore always potentially contains space for alternative acts and alternative intentions which are not yet articulated as a social institution or even project'.

50 See the valuable discussion in Burke (1987), e.g. p. 220: 'The notion of cultural hegemony might well be useful . . . if we could only identify the places and times when hegemony (rather than independence or coercion) was in operation, and the conditions which give rise to this situation.' Burke's justified optimism about integrating social and cultural domains contrasts refreshingly with Darnton's (1986) recent pessimism.

51 The problem – and patrician hegemony's hat-trick – has been the identification of democracy as vulgar, to the extent of convincing the disenfranchised themselves.

52 See the excellent discussion in Howe (1984) ch. 3.

53 The quotation is from Milan Kundera, The *Unbearable Lightness of Being* (London: Faber, 1984).

54 Allowing for the contemporary sexist language, this was matchlessly formulated by Marx (1984) 10: 'Men make their own history, but they do not make it just as they please; they do not make it under circumstances

188

chosen by themselves, but under circumstances directly encountered, given and transmitted from the past.' Against my interpretation here, there is of course a more pessimistic view, recently expressed by Seamus Heaney: 'As if the eddy could reform the pool.'

55 Milan Kundera, *The Art of the Novel* (London: Faber, 1986), pp. 7, 164. This approach is what I have called critical or (after Laclau and Mouffe) democratic pluralism. It entails a certain kind of relativism, with implications not hostile to criticism (as is commonly misunderstood) but to certain kinds of criticism, e.g. transcendental. As the reader will probably realize by now, I see relativism as a desideratum, not a problem. For able explication and defence of the real thing, as distinct from various straw bogey-men, see Feyerabend (1978) 79–86 and (1987), Fish (1980), Collins (1981), Rorty (1986), and Kuhn's 1987 Sherman memorial lectures; but also Hirst (1985). On the teleological realism of some Marxist historians, see Pocock (1980, 1984) and Clark (1986). A robust social history of ideas à *la* Thompson et al. should actually welcome Clark's attack on teleology and anachronism, and re-articulate it in terms of a non-authoritarian 'democratic pluralism' (Laclau and Mouffe, 1986) and 'democratic epistemology' (from the excellent discussion in Barrow (1986) 146–61). Such a move would not only help neutralize much of the intellectual legitimacy of current conservatism, but would also represent a real advance on the left's atavistic attachment to such anti-democratic residues in its own tradition, and the false sense of security and superiority that these encourage.

Bibliography

Primary Sources

A[llen], J[ohn], *Judicial Astrologers Totally Routed* . . . *Or a Brief Discourse, wherein is clearly manifested, That Divining by the Stars hath no solid Foundation* (London, 1659).

——, *Several Cases of Conscience, concerning Astrologie, And seekers unto Astrologers, Answered* . . . (London, 1659).

Allen, Thomas, *Claudii Ptolemai pelusiensis de astrorum judiciis, aut ut vulgo vocant Quadripartite constructionis* . . . Ash. MS 388, n.d.

Andrews, Henry, in *Moore's Vox Stellarum* (*c.*1783–1820).

Aristotle's Book of Problems . . . .*Wherein is contain'd divers, Questions and Answers touching the state of man's body* . . . (London, ?1710 and throughout the eighteenth century).

Aristotle's Masterpiece (London, 1684, ?1710 and throughout the eighteenth century).

Articles to be Enquired of . . . in the Ordinary Visitation of . . . Charles, Lord Bishop of Norwich (Norwich, 1716; Bodl. GP 1523 [27*, 28]).

Ashmole, Elias, *Fasciculus Chemicus* . . . (London, 1650).

——, *Theatrum Chemicum Britannicum* . . . (London, 1652).

——, *Memoirs of the life of that learned Antiquary Elias Ashmole, Esq.; Drawn up by himself by way of Diary* . . . (London, 1717).

——, *The Lives of those eminent antiquaries Elias Ashmole, Esquire, and Mr. William Lilly, written by themselves* . . . (London, 1774).

——, *The Diary and Will of Elias Ashmole, edited and extended from the Original Manuscripts by R. T. Gunther* (Oxford: for the Subscribers, 1927).

——, *Elias Ashmole (1617–1692): His Autobiographical and Historical Notes, His Corresspondence, and Other Contemporary Sources Relating to his Life and Work*, ed. C. H. Josten (Oxford: Clarendon Press, 1966; 5 vols).

Astrologus, *The Celestial Telegraph, or Almanack of the People* . . . (London, 1796).

Atkinson, J. C., *Forty Years in a Moorland Parish* . . . (London, 1891).

Atwell, George, *An Apology, or, Defence of the Divine Art of Natural Astrology* . . . (London, 1660).

Aubrey, John, *Brief Lives, Chiefly of Contemporaries* . . . *between the years 1669 and 1696* . . ., (ed. Andrew Clark, Oxford, 1898; 2 vols).

——, 'Collectio geniturary' (1669–96; Aubrey MS. 23).

Bacon, Francis, *The Works of Francis Bacon* . . ., ed. James Spedding, Robert Leslie Ellis, and Douglas D. Heath (London, 1857–74; 14 vols).

Ball, Richard, *An Astrolo-Physical Compendium. Or a Brief Introduction to Astrology* (London, 1697; 2nd edn (1723) published as *Astrology Improv'd: or, A Compendium of the Whole Art of that Most Noble Science)*.

——, *A Warning to Europe: Being Astrological Predictions on the Conjunction of Saturn, Jupiter, and Mars* . . . (London, 1722).

Bamford, Samuel, *Passages in the Life of a Radical* (London and Manchester, 1841; 2 vols).

Barnes, Joshua, *An Elegy on the Death of the Reverend Dr John Goad* (1689; Wood MS 429 [44]).

Bayle, Pierre, *Miscellaneous Reflections occasion'd by the Comet which appear'd in December 1680* (London, 1708; first published as *Lettre* . . . *où il est prouvé* . . . *que les comètes ne sont point le présage d'aucun malheur* . . . [Cologne, 1682]).

——, *An Historical and Critical Dictionary* . . . (London, 1710; first published in Rotterdam, 1697; 4 vols).

[Beale, John], 'The Copy of a Letter from Somerset . . .' *Phil. Trans.* 7 (1673) 5141–12.

Beaumont, George, *Fixed Stars: or, An Analyzation and Refutation of Astrology* . . . (Norwich, 1803).

Beckworth, Barnaby, *A Warning Piece for London, or, A Completion of Several Prophecies* . . . *in these our days* . . . (London, 1733).

Beddoes, Thomas (Ed.), *Contributions to Physical and Medical Knowledge, principally from the West of England* (Bristol, 1799).

Beetenson, William, *Prodromus Astrologicus: Being an Astrological Discourse, of the Effects of the Great Eclipse of the Sun* . . . (London, 1722).

Birch, Thomas, *The History of the Royal Society of London* (London, 1756–7; 4 vols).

Blackwel, James, *The Nativity of Mr. William Lilly, Astrologically Reformed* . . . (London, 1660).

Blagrave, Joseph, *Astrological Practice of Physick* (?Reading, 1671; the 2nd edn published as Blagrave's Astrological Practice of Physick [1672]).

——, *An Introduction to Astrology* . . ., ed. Obadiah Blagrave (London, 1682).

Boehme, Jacob, *Aurora. That is, the Day-Spring* . . ., transl. John Sparrow (London, 1656).

——, *Several Treatises of Jacob Behme* . . ., transl. John Sparrow (London, 1661).

——, *The Remainder of Books Written by Jacob Behme* . . ., transl. John Sparrow (London, 1662).

——, *The Works of Jacob Behme, The Teutonic Philosopher*, ed. and transl. William Law, with George Ward and Thomas Langcake (London, 1764–81; 4 vols).

[——], *A Compendious View of the Grounds of the Teutonick Philosophy*, ed. John Pordage (London, 1770).

Book of the Prodigies, or Book of Wonders, or Mirabilis Annus (London, 1662).

Booker, John, *A Bloudy Irish Almanack; or Rebellious Bloody Ireland Discovered* . . . (London, 1646).

——, in miscellaneous almanacs from 1631, published 1659–67 under the title of *Telescopium Uranicum*.

Bourne, Benjamin, *The Description and Confutation of the Mysticall Anti-Christ, the Familists, &c* . . . . (London, 1646).

Boyle, Robert, *The Works of the Honourable Robert Boyle* . . ., ed. Thomas Birch (London, 1744; 5 vols; 2nd edn in 6 vols published in 1772).

——, [on astrology] Boyle Papers, XIX, 298; [on judicial astrology] Boyle Papers, VIII, 204–5.

Boyse, Samuel, *The New Pantheon: or, Fabulous History of the Heathen Gods* . . . *&c* . . . *To which is subjoin'd an appendix treating of their Astrology* . . ., ed. and rev. William Cooke (Salisbury, 1760?).

Brand, John, *Observations on Popular Antiquities, including the Whole of Mr. [Henry] Bourne's Antiquitates Vulgares [1725]* (London, 1776; 3 vols; another edition as *Brand's Popular Antiquities of Great Britain* . . ., ed. by W. Carew Hazlitt (London, 1905)).

Brayne, John, *Astrologie Proved to be the Old Doctrine of Demons* . . . (London, 1653).

A Brief Answer to Six Syllogistical Arguments Brought by Mr. Clark . . . *Against Astrologers and Astrologie* (London, 1660).

Brookes, Richard, *The General Practice of Physic* . . ., 6th edn (London, 1771; 2 vols).

Brothers, Richard, *A Revealed Knowledge of the Prophecies and Times* (London, 1795).

——, *A Description of Jerusalem: Its Houses and Streets* (London, 1801 [1802]). published 1707–8 in 3 vols).

Burton, Robert, *The Anatomy of Melancholy* . . . (New York: Vintage, 1977; first published (in London) 1621, with further editions throughout the seventeenth century).

Butler, John, *[Christologia], or a Brief but True Account of the Certain Year,*

Moneth, Day and Minute of the Birth of Jesus Christ (London, 1671).

——, *[Astrologia]*, or, *The most Sacred and Divine Science of Astrology* . . . (London, 1680).

Butler, Samuel, *Hudibras* (London, 1663 [1662], 1664 [1663], and 1678; 3 vols).

Calendar of State Papers, Domestic Series (London, vols for 1661–2; 1663–4; 1666–67; 1679–80).

Calvin, Jean, *An admonicion against Astrology iudiciall and other curiosities, that raigne now in the world* . . . translated into English by G[odred] G[ylby] (London, 1651; first published in Latin in 1549, also in French in 1566).

Carleton, George, *[Astrologomania:] The Madness of Astrologers* . . . (London, 1651).

Carpenter, Richard, *Astrology Proved Harmless, Useful, Pious* . . . (London, 1657).

Casaubon, Meric, *A Treatise Concerning Enthusiasme* . . . (London, 1655).

——, *Of Credulity and Incredulity, in things Natural, Civil, and Divine* (London, 1668).

Catastrophe Mundi: or Merlin Reviv'd (London, 1683).

Chamber, John, *A Treatise Against Iudiciall Astrologie* . . . (London, 1601).

Chambers, Ephraim, *Cyclopaedia: or, an Universal Dictionary of Arts and Sciences* . . . (London, 1728; 3 vols).

Charles, Lord Bishop of Norwich, *Articles to be Enquired of . . . in the Ordinary Visitation of Charles* . . .(Norwich, 1716).

Childrey, Joshua, *Indago Astrologica: or, A brief and modest Enquiry into some Principal Points of Astrology* . . . (London, 1652).

——, *Syzygiasticon Instauratum: or, an Almanack and Ephemeris for the Year . . . 1654* (London, 1653).

——, Letter (to Samuel Hartlib) of 14 August 1655 (Sloane MS 427, ff. 69–82).

——, *Britannia Baconia: or, The Natural Rarities of England, Scotland, and Wales* . . . (London, 1660).

——, Letter (to Henry Oldenburg) of 14 April 1669, in Oldenburg (1968) 488.

——, Letter (to Seth Ward, along with a reply from John Wallis), of 10 October, 1670, in *Phil. Trans.* 5 (1670) 2061–74.

Clare, John, *A Shepherd's Calendar; with Village Stories and other Poems* (London, 1827).

[Clarendon, The Earl of], *His Majesties Most Gracious Speech, Together with the Lord Chancellor's, to the Two Houses of Parliament; on Thursday the Thirteenth of Spetember, 1660* (London, 1660).

Clarke, Samuel, [1599–1683] *Medulla Theologiae* (London, 1659).

Clarke, Samuel, [1675–1729] *Sermons* . . . (London, 1730; 10 vols).

Clarkson [or Claxton], Laurence, *The Lost Sheep Found: or, the Prodigal returned to his Father's house* . . . (London, [1660]).

Coley, Henry, *Clavis Astrologiae elimata; or A Key to the Whole Art of Astrologie* (London, 1669).

Congreve, William, *Love for Love [a play]* (London, 1695).

Cowley, J., *Discourse on Comets* (London, 1757).

C.P., *The Sheepherd's New Kalender* . . . (London, 1700).

Culpeper, Nicholas, *A Physical Directory: or a Translation of the Dispensatory made by the College of Physicians of London* . . . (1649).

——, *Semiotica Urania* . . . (London, 1651).

——, *Catastrophe Magnatum: or, The Fall of Monarchie* . . . (London, 1652).

——, *The English Physitian, or an Astrologo-Physical Discussion of the Vulgar Herbs of this Nation* . . . (London, 1652b; further editions throughout the seventeenth and eighteenth centuries).

——, *Pharmacopoeia Londinensis: or the London dispensatory* . . . (London, 1653).

——, *Opus Astrologicum* . . . (London, 1654a).

——, *A New Method of Physick* . . . (London, 1654b).

——, *An Ephemeris* . . . [an almanac published annually 1651–6].

——, *Astrological Judgement of Diseases* (London, 1665).

Dariot, Claude, *A breefe and most easie Introduction to the Astrological Judgement of the Starres* . . . ., transl. Fabian Wither (London, 1583).

——, *Dariotus Redivivus: Or a briefe Introduction Conducing to the Judgement of the Stars* . . . .*much enlarged* . . . *by Nathaniel Spark* . . . (London, 1653).

Davy, Humphry, 'An Essay on Heat, Light, and the Combinations of Light', pp. 5–205 in Beddoes (Ed.) (1799).

[Defoe, Daniel], *A Journal of the Plague Year* . . . (London, 1722).

——, *A System of Magicke; or A History of the Black Art* (London, 1727).

Denam, M. Aislabie, *Folk-lore; or, Manners and Customs of the North of England* (London, 1850).

Derham, William, *Astro-Theology, or a Demonstration of the Being and Attributes of God, from a Survey of the Heavens* . . . (London, 1715; further editions throughout the eighteenth century).

Diderot, Denis, and D'Alembert, Jean, *Encyclopédie, ou Dictionnaire Raisonnée des Sciences, des Arts et des Métiers* . . . (Paris, 1751–75; 17 vols).

The Dreadful Effects of Going to Conjurers . . . (London, 1708).

Dryden, John, *The Works of John Dryden* (Edinburgh, 1821; 18 vols).

——, *The Letters of John Dryden*, ed. Charles E. Ward (Durham, NC: University of North Carolina Press, 1942).

——, *Annus Mirabilis: The Year of Wonders, 1666. An Historical Poem* . . . (London, 1667).

Dyer, T. F. Thistleton, *British Popular Customs, Present and Past* . . . (London, 1876).

——, *English Folklore* (London, 1880).

Edlin [or Edlyn], Richard, *Observationes astrologiae* . . . (London, 1659).

——, *Prae-Nuncius Sydereus: an Astrological Treatise of the effects of the great conjunction of . . . Oct. the Xth 1663* . . . (London, 1664).

[Edwards, John], *Cometomantia. A Discourse of Comets* . . . .*Wherein is also inserted an Essay of Judiciary Astrology* (London, [1684]).

Eland, William, *A Tutor to Astrology, or Astrology made easie* . . ., 'the 7th edition' (London, 1694).

——, *Eland's Tutor to Astrology* . . . .*The tenth edition. Corrected* . . . *and enlarged* . . . *by George Parker* (London, 1704).

An Elegy Upon the Death of Mr. William Lilly, The Astrologer (London, 1681; printed for Obadiah Blagrave).

Encyclopaedia Britannica . . . 1st edn (Edinburgh, 1771; 3 vols).

Evelyn, John, *The Diary of John Evelyn* . . ., ed. William Bray (London: Bickers and Son, 1879; 4 vols).

Farmer, Ralph, *A Sermon Preached Against Astrology and Astrologers* (Bristol, 1651).

Fart-inando, *The Asses of Great Britain* . . . *by Fart-inando, a modern political astrologer* . . . (London, [1762]).

Ferguson, James, *An Idea of the Material Universe, Deduced from a Survey of the Solar System* (London, 1754).

Ferrier, Auger, *A Learned Astronomical Discourse of the Judgement of Nativities*, transl. Thomas Kelway; (London, 1593; first published as *Jugements Astronomiques sur les Nativitéz* [1550]; a second English edition in 1642).

Fitzsmith, Richard, *Syzygiasticon Instauratum: or, an Almanack and Ephemeris for . . . 1654* (London, 1654).

Flammarion, Camille, *The Marvels of the Heavens*, transl. Norman Lockyer) (London, 1870).

Flamsteed, John, 'Hecker. His large Ephemeris . . . ' (1674) [see Hunter (1987)].

——, Letter (to Richard Towneley) of 11 May 1677 (Royal Society MS LIX.c.10).

——, 'An Exact Account of the Three Late Conjunctions of Saturn and Jupiter', *Phil. Trans.* 13 (1683) 244–58.

Fletcher, J. S., *Recollections of a Yorkshire Village* (London, 1910).

Further Testimonies of the Authenticity of the Prophecies of Richard Brothers, astrologically accounted for . . . (London, 1795).

Gadbury, John, *Philastrogus Knavery Epitomized* . . . (London, 1652).

——, *Animal Cornatum; or . . . A brief method of the grounds of Astrology* . . . (London, 1654).

——, *Genethlialogia, or The Doctrine of Nativities, and the Doctrine of Horary Questions, Astrologically Handled* (London, 1658).

——, *Nuncius Astrologicus* . . . (London, 1659).

——, *Natura Prodigiorum: or, A Discourse Touching the Nature of Prodigies* . . . (London, 1660).

——, *Britain's Royal Star* . . . (London, 1661).

——, *Collectio Geniturarum, or A Collection of Nativities* . . . (London, 1662).

——, *Dies Novissimus: or, Dooms-Day Not so Near as Dreaded* . . . (London, 1664).

——, *London's Deliverance Predicted* . . . (London, 1665a).

——, *De Cometis, or A Discourse on the Natures and Effects of Comets* . . . (London, 1665b).

——, *A Brief Relation of the Life and Death of* . . . *Mr. Vincent Wing* . . . (London, 1669).

——, *The West India or Jamaica Almanack* . . . (London, 1674).

——, *Obsequium Rationabile, or A Reasonable Service performed for the Coelestial sign Scorpio* . . . (London, 1675).

——, *Ephemerides of the Celestial Motions and Aspects* . . . *for XX Years* . . . (London, 1680).

——, *Cardines Coeli* . . . (London, 1684).

——, *Nauticum Astrologicum, or, The Astrological Seaman* . . . (London, 1689).

——, in *[Ephemeris.]* or, *A Diary Astronomical and Astrological* . . . an almanac published 1659–1703).

——, [examined by the Cabinet Council, twice, on 11 June, 1690] *Historical Manuscripts Commission* 71 (1957) 379.

Gassendi, Pierre, *The Vanity of Judiciary Astrology, or Divination by the Stars* . . ., (transl. by 'a gentleman', London, 1659).

Gassendus's Arguments Against Astrologie . . . *Retorted and Refuted* . . . (London, 1660).

Gataker, Thomas, *Annotations upon all the Books of the Old and New Testaments* . . ., 2nd edn (London, 1651; second edition).

——, *His Vindication* . . . *Against the Scurrilous Aspersions of* . . . *Mr William Lillie* . . . (London, 1653).

——, *A Discours Apologetical, wherein Lilies* . . . *Lies* . . . *are clearly laid open* . . . (London, 1654).

Gaule, John, *Pus-mantia. The Mag-astro-mancer, or the Magicall-Astrologicall Diviner Posed and Puzzled* . . . (London, 1652).

Gell, Robert, *Stella Nova, A New Starre Leading Wisemen unto Christ* . . . (London, 1649).

——, . . . *Or a Sermon Touching Gods government of the World by Angels* . . . (London, 1650).

——, *The New Jerusalem* . . . (London, 1652).

Gell's Remaines . . . (London, 1676).

Geree, John, *Astrologo-Mastix, or a Discovery of the Vanity and Iniquity of Judiciall Astrology* . . . (London, 1646).

Gibson, Richard, *Flagellum Placidianum, or A Whip for Placidianism* . . . (Gosport, 1711).

Goad, John, ' . . . An Advent Sermon . . . ' (1663; Bodleian Pamphlet C.118.(17)).

——, 'A Sermon Treating of the Tryall of all Things by the Holy Scriptures . . .' (1664; Bodl. Vet. A3.e.173).

——, *Astro-Meteorologica, or Aphorisms and Discourses of the Bodies Coelestial, their Natures and Influences* . . . *Collected from the Observation* . . . *of above Thirty Years* (London, 1686).

——, *Astro-meteorologia sana* . . . (London, 1690).

——, Weather Diary (Sloane MS 1731, A, ff. 31–2; and Ashmolean MS. 368).

Godson, Robert, *Astrologia Reformata: A Reformation of the Prognostical Part of Astronomy, Vulgarly Termed Astrology* . . . (London, 1697).

Gough, Richard, *The History of Myddle, ed. David Hey* (Harmondsworth: Penguin, 1981; first published as *Human Nature Displayed in the History of Myddle* [1834]).

Graunt, John, *Natural and Political Observations Made on the Bills of Mortality* . . . (London, 1662).

Gregory, David, 'De astrologia . . .' (1686; Christ Church College [Oxford] MS 133, ff. 47–50).

——, *The Elements of Physical and Geometrical Astronomy* . . . (London, 1726; first published as *Astronomiae physicae et geometriae elementa* . . . in 1702]).

Hall, Thomas, *Histrio-Mastix. A Whip for Webster* . . . (London, 1654).

Halley, Edmond, *Correspondence and Papers of Edmond Halley*, ed. E. F. MacPike (Oxford: Clarendon Press, 1932).

[Hammond, Anthony], *Solitudinus Munus: or, Hints for Thinking* (London, 1721).

Harland, John and Wilkinson, T. T., *Lancashire Folk-lore* . . . (London, 1867).

Harley, Timothy, *Moonlore* (London, 1885).

Harris, John, *Lexicon Technicum: or an Universal English Dictionary of Arts and Science* . . . (London, 1704, 1710; 2 vol.).

Hazlitt, William, *The Complete Works of William Hazlitt*, ed. P. P. Howe (London, 1931; 21 vols).

Hearne, Thomas, *Remarks and Collections of Thomas Hearne*, ed. C. E. Dobb et al. (Oxford: Oxford University Press, 1885–1921; 11 vols).

Helmont, John Baptista van, *Oriatrike, or, Physick Refined* . . ., transl. J[ohn] C[handler] (London, 1662; first published as *Ortus Medicinae* . . . (Amsterdam, 1648)).

Henderson, May Sturge, *Three Centuries in North Oxfordshire* . . . (London: Edward Arnold, 1902).

Henderson, William, *Notes on the Folk-lore of the Northern Counties of England and the Borders* . . . (London, 1866; another edition 1879).

Heydon, Sir Christopher, *A Defence of Iudiciall Astrologie* . . . (Cambridge, 1603).

——, *An Astrological Discourse* . . . (London, 1650).

Heydon, Christopher Jr. [sic], *The New Astrology* . . . (London, 1785).

Heydon, John, *The Rosie Crucian. Infallible Axiomata . . . to know all things past, present, and to come* . . . (London, 1660).

——, *The Holy Guide . . . whereunto is added, A Bar to Stop Thomas Streete* . . . (London, 1662a).

——, *The Harmony of the World* . . . (London, 1662b).

——, *Theomagia, or The Temple of Wisdome* . . . (London, 1664).

Hill, John, *Urania: or, A Compleat View of the Heavens* . . . (London, 1754).

Hobbs, William, *The Earth Generated and Anatomized* . . . ed. Roy Porter (London: British Museum, 1981).

Holwell, John, *Catastrophe Mundi: or, Europe's Many Mutations* . . . (London, 1682).

Homes, Nathanael, *Plain Dealing* . . . (London, 1652).

Hone, William, *The Yearbook of Daily Recreation and Information* . . . (London, 1832).

Hooke, Robert, *The Diary of Robert Hooke*, ed. H. W. Robinson and Walter Adams (London: Taylor & French, 1935).

Hunt, William, *The Demonstration of Astrology. Or, a brief Discourse, proving the Influence of the Sun, Moon, and Stars* . . . (London, 1696).

Irish, David, *Animadversio Astrologica: or, A Discourse Touching Astrology* . . . (London, 1701).

J. B., *The Blazing Star; or, A Discourse of Comets* . . . (London, 1665).

Jeake, Samuel, [see Hunter and Gregory (1987)].

Johnson, H., *Anti-Merlinus: or a Confutation of Mr. William Lillies Predictions* . . . (London, 1648).

Johnson, Samuel, *A Dictionary of the English Language* (London, 1755; 2 vols; further editions up to 1827).

Josselin, Ralph, *The Diary of the Reverend Ralph Josselin, 1616–1683* ed. Alan MacFarlane (London: Oxford University Press, 1976).

Kepler, Johannes, [see Field (1984; 1987) and Brackenridge (1979)].

Key, R. ['Student in Astrology'], *A Theory of New Philosophy* . . . (London, ?1750).

Kirby, Richard, and Bishop, John, *The Marrow of Astrology* . . . (London, 1687; 2nd edition (1688) in Bishop's name only).

Knight, Charles, *Passages of a Working Life During Half a Century* . . . (London, 1864–65; 2 vols).

Lambert, J. H., *Cosmological Letters on the Arrangement of the World-Edifice*, ed. Stanley L. Jaki (Edinburgh: Scottish Academic Press, 1976; first published in German in 1761).

Latham, Charlotte, 'Some West Sussex Superstitions Lingering in 1868', *Folklore Society [Records]* 1 (1868) 7–61.

Leadbetter, Charles, *A Treatise of Eclipses for Twenty-Six Years* . . . (London, 1717).

Le Neve, Geoffrey, 'Vindicta Astrologicae Judiciariae' (Ash. MS. 418).

Lilies Banquet: or, the Star-Gazers Feast . . . (London, 1653).

Lilly, William, *A Prophecy of the White King and Dreadfull Dead-man Explaned* (London, 1644).

——, *The Starry Messenger, or, an interpretation of that strange apparition of three Suns* . . . (London, 1645).

——, *Christian Astrology, modestly treated of in three Books* . . . (London: Regulus, 1985; with afterwords by P. Curry and G. Cornelius; first published 1647, with a 2nd edn in 1659).

——, *Astrologicall Predictions of the Occurrences in England* . . . (London, 1648).

——, *Monarchy or No Monarchy in England* . . . (London, 1651).

——, *Annus Tenebrosus* . . . (London, 1652).

——, in *Merlinus Anglicus Junior*, an almanac for 1644; *Anglicus* . . ., almanacs for 1645 and 1646; and *Merlini Anglici Ephemeris* . . ., an almanac appearing annually 1647–82, and 1683–5 posthumously.

——, *The Last of the Astrologers* . . . (London, 1974; first published as *Mr. William Lilly's History of his Life and Times from the year 1602 to 1681* . . . [1715], with further editions in the eighteenth and nineteenth centuries).

Long, Roger, *Astronomy, in Five Books* (Cambridge, 1742–64; 2 vols).

Lovett, William, *Life and Struggles of* . . . *In his Pursuit of Bread, Knowledge and Freedom* . . . (London, 1876).

L. P., *The Astrologer's Bugg-beare* . . . (London, 1652).

[Man, John], *The Stranger in Reading* . . . (Reading, 1810).

Maternus, Firmicus, *Ancient Astrology: Theory and Practice. The Mathesis of Firmicus Maternus*, transl. Jean Rhys Bram (Park Ridge, NJ: Noyes Press, 1975).

Mead, Richard, *A Treatise concerning the Influence of the Sun and Moon upon Human Bodies, and the Diseases thereby produced* . . . (London, 1748; first published as *De Imperio Solis ac Lunae in corpora humana* . . . [1704]).

Melton, John, *Astrologaster, or, The Figure Caster* . . . (London, 1620).

Mensforth, George, *The Young Student's Guide to Astrology* . . . (London, 1785).

Mercator, Nicholaus, 'Astrologia rationalis, argumentis solidus explorata' (Shirburn MS 180.F.34).

Merrifield, John, *Catastasis Mundi: or, the True State . . . of Christendom . . .* (London, 1684).

Middleton, John, *Practical Astrology . . .* (London, 1679).

Mirabilis Annus, or the Year of Prodigies and Wonders . . . (London, 1661).

Mirabilis Annus Secundus . . . (London, 1662).

Moore, Francis, in *Vox Stellarum . . .*, an alamanac published annually 1699–1714, and posthumously as *Moore's Vox Stellarum/Almanac* up to the present.

More, Henry, *An Explanation of the Grand Mystery of Godliness . . .* (Cambridge, 1661).

——, *Tetractys Anti-Astrologica . . .* (London, 1681; reprinted from his 1661).

Morin, Jean-Baptiste, *Astrologia Gallica principiis et rationibus propis stabilita, atque in XXVI Libros distributa* (The Hague, 1661; Book 21 has been translated by R.S. Baldwin and published as *The Morinus System of Horoscope Interpretation* (Washington DC: AFA, 1974)).

——, 'The Cabal of the Twelve Houses Astrological, from Morinus', transl. George Wharton; published in his almanac of 1659; republished as pp. 189–207 in Wharton (1683); first published as 'Astrologicorum domorum cabala . . .' (Paris, 1628).

——, 'Teaching how Astrology may be restored: from Morinus, *viz.*', transl. George Wharton; pp. 184–9 in Wharton (1683); first published as 'Ad australes et boreales astrologus pro astrologia restituenda epistolae' (Paris, 1628).

——, '. . . the Astrological axioms and theorems of Morinus', transl. Henry Coley, in his almanac of 1672, from ch. 29 of Morin (1661).

Moxon, Joseph, *A Tutor to Astronomie and Geographie . . .* (London, 1659).

Napier, John, *Memoirs of John Napier of Merchiston . . .*, ed. M. Napier (Edinburgh, 1834).

Naylor, M. J., *The Inantity [sic] and Mischief of Vulgar Superstitions* (Cambridge, 1795).

No Cheat nor Meer Pretended Fortune-teller, but an honest and faithful Student in Astrology . . . (London, ?1712).

Oldenburg, Henry, *The Correspondence of Henry Oldenburg*, ed. A. Rupert and Mary Boas Hall (Madison, WI: University of Wisconsin Press, 1968; 8 vols).

Parker, George, in *Mercurius Anglicanus . . .*, an almanac, published annually 1690–8).

——, *An Ephemeris . . . Heliocentric and Geocentric . . .*, published annually 1695–9, and thenceforth appearing as *Parker's Ephemeris* until 1781 (posthumously after 1743).

——, *The West-India Almanack . . .* (London, 1719).

Parker, Samuel, *A Free and Impartial Censure of the Platonicke Philosophie* (Oxford, 1666).

Partridge, John, *Vox Lunaris* . . . (London, 1679).

——, *Mene Mene* . . . (London, 1688).

——, *Mene Mene Tekel Upharshin* . . . (London, 1689a).

——, *Remarkable Predictions of that Great Prophet Michael Nostradamus* . . . (?London, 1689b).

——, *Opus Reformatum: or, A Treatise of Astrology. In which the Common Errors of that Art are Modestly Exposed and Rejected. With an Essay towards the Reviving the True and Ancient Method laid down . . . by the Great Ptolemy* . . . (London, 1693).

——, *Defectio Geniturarum: Being an Essay toward the Reviving and Proving the True Old Principles of Astrology* . . . (London, 1697a).

——, *Flagitiosis Mercurius Flagellatus* . . . (London, 1697b).

?—— Mr. Patridge's [sic] Judgement and Opinion of this Frost . . . (London, 1709).

——, in *Mercurius Coelestis*, an almanac for 1681–2); *Mercurius Redivivus*, an almanac for 1683–7; and *Molinus Liberatus*, an almanac 1690–1709, 1713–14, and henceforth posthumously into the late eighteenth century.

——, *The Last Wills and Testaments of* . . . (London, 1716; dated 3 December 1714).

Partridge and Flamsted's New and Well Experienced Fortune Book . . . (London, ?1750 with further editions in the late eighteenth century).

Peacock, M., 'Folklore and Legends of Lincolnshire' (MS in Folklore Society Library, n.d.).

——, in *Notes and Queries*, (1856) 415.

Penny, James Alpass, *Folklore Around Horncastle* (Horncastle: W. K. Morton, 1915).

Penseyre, Samuel, *A New Guide to Astrology* . . . (London, 1726).

Pepys, Samuel, *The Diary of Samuel Pepys*, ed. Robert Latham and William Matthews (London: Bell & Hyman, 1970–83; 11 vols).

Petty, William, *The Advice of W.P. to Mr. Samuel Hartlib For the Advancement of some particular Parts of Learning* (London, 1647).

Phillips, R., *The Celestial Science of Astrology Vindicated* . . . (London, 1785).

Placidus de Titis, *Physiomathematica* . . . (Milan, 1646–50).

——, *Primum Mobile* (London: ISCWA, 1983; a reprint of the first English translation by John Cooper (London, 1814); first published as *Tabulae Primi Mobilis* . . . [Padua, 1657]).

——, *An appendix concerning Part of Fortune* . . ., transl. John Whalley (Dublin, 1701; appended to his edition of 1701).

——, *Astronomy and Elementary Philosophy translated from the Latin of Placidus de Titus*, transl. Manoah Sibly (London, 1789a).

——, *A Collection of thirty remarkable nativities to illustrate the canons* . . . *of Placidus de Titus* . . ., transl. Manoah Sibly (London, 1789b).

201

——, *Supplement to Placidus de Titus* . . ., transl. Manoah Sibly (London, 1789c).

Plot, Robert, Letter to Martin Lister, 23 March 1684, *Phil. Trans.* 15 (1684) 930–31.

Pool, John, *Country Astrology* . . . (London, 1650; actually a translation of Ferrier's *Learned Astronomical Discourse* (1642), a point which I owe to J. Halbronn).

Prescot, Bartholomew, *The Inverted Scheme of Copernicus* . . . (Liverpool, 1822).

Ptolemy, Claudius, *Ptolemy's Quadripartite* . . ., transl. John Whalley (London, 1701; 2nd edition in 1786, 'revised, corrected and improved' by Manoah Sibly).

——, *Tetrabiblos*, transl. and ed. F. E. Robbins (London and Cambridge, MA: Heinemann, 1940; there are also earlier translations by James Wilson (1820) and J. M. Ashmand (1822).

Purslow, Norris, Diary 1673–1737 (Wellcome MS 4021).

Ramesey, William, *Lux Veritatus, or, Christian Judicial Astrology Vindicated, and Demonology Confuted* . . . (London, 1650).

——, *Vox Stellarum* . . . (London, 1652).

——, *Astrologia Restaurata* . . . (London, 1653).

Raunce, John, *L'Astrologia accusata partiter et condemnata* . . . (London, 1650a).

——, *A Brief Declaration against Judicial Astrologie* . . . (London, 1650b).

Ray, John, *Miscellaneous Discourses Concerning the Dissolution and Changes of the World* . . . (London, 1713; first published in 1692).

Reeve, Edmund, *The New Jerusalem* . . . (London, 1651).

Reid, William Hamilton, *The Rise and Dissolution of the Infidel Societies in this Metropolis* . . . (London, 1800).

Rohault, Jacques, *A System of Natural Philosophy* . . ., transl. John Clarke (London, 1723; 2 vols).

Rowland, William, *Judiciall Astrologie, Judicially Condemned* (London, 1651).

Salmon, William, *Synopsis Medicinae: Or, A Compendium of* . . . *Physick* . . . (London, 1671).

——, *Horae Mathematicae, seu Urania. The Soul of Astrology* . . . (London, 1679).

——, *Iatricia seu Praxis Medendi* . . . (London, 1681; published in fortnightly instalments).

——, in *The London Almanack* . . ., published annually 1692–1706.

Saunder, Richard, in *Apollo Anglicanus* . . ., an almanac, published annually 1684–1736.

Saunders, Richard, *The Astrological Judgement and Practice of Physick* . . . (London, 1677).

——, in *Apollo Anglicanus* . . ., an almanac, published annually 1654–75.

Season, Henry, in *Speculum Anni* . . ., an almanac, published annually 1733–75.

Shakerley, Jeremy, letters to William Lilly, 1647–50 (Ash. MS 423 II, ff. 111–28.

——, *The Anatomy of Urania Practica* . . . (London, 1649).

——, letter to John Matteson, 5 March 5 1648–9, *Historical Manuscripts Commission* 8 (1913) 61.

The Shepherd's Kalender (London, 1503; translated from French; at least eighteen editions till 1650).

Sibly, Ebenezer, *Uranoscopia, or the pure language of the Stars* . . . (London, c.1780).

——, . . . *A New and Complete Illustration of the Celestial Science of Astrology* . . . (London, 1784–92; 4 vols; further editions, some entitled *A New and Complete Illustration of the Occult Sciences*, published up to 1826).

——, *Culpeper's English Physician: and Complete Herbal* . . . (London, 1789; at least thirteen further editions up to 1813).

Society of Astrologers, Letter to B. Whitelocke, 24 April 1650 (Ash. MS. 423, ff. 168–9).

Southcott, Joanna, *Divine and Spiritual Letters of Prophecies* . . . (London, 1801).

[Southey, Robert], *Letters from England* . . . (London, 1807; 3 vols).

——, *The Life and Correspondence of the late Robert Southey*, ed. C. C. Southey (London, 1849–50; 6 vols).

Spittlehouse, John, *Rome Ruined by Whitehall* (London, 1650).

Sprat, Thomas, *The History of the Royal Society of London* . . . (London, 1667).

Stokes, Edward, *The Wiltshire Rant* . . . (London, 1652).

Streete, Thomas, *A Double Ephemeris* . . . (London, 1653).

——, *Astronomia Carolina, A New Theorie of the Coelestial Motions* . . . (London, 1661; another edition ed. E. Halley in 1710).

——, *An Appendix to Astronomia Carolina* . . . (London, 1664).

——, *A Compleat Ephemeris* . . . (London, 1684).

Stukeley, William, *The Family Memoirs of the Reverend William Stukeley*, W. C. Lukis (London, 1882–3; 2 vols; vols 73 and 76 of the *Publications of the Surtees Society*).

——, Wellcome MS 4729 and BL MSS 50148, 50151.

The Surprising Monument . . . *Together with Dr Flamsted, Mr. Halley, Mr. Wiston, Dr Partridge, and several other Astrologers' opinions* . . . (?London, ?1715).

Swadlin, Thomas, *Divinity No Enemy to Astrology* . . . (London, 1653).

Swan, John, *Signa Coeli; the Signs of Heaven* . . . (London, 1652).

[Swift, Jonathan], *Esquire Bickerstaff's Most Strange and Wonderful Predictions* . . . (London, 1708a).

[——], *The Accomplishment of the First of Mr. Bickerstaff's Predictions* . . . (London, 1708b).

[——], *A Vindication of Isaac Bickerstaff* . . . (London, 1709).

?[——], *Bickerstaff's Almanack: or, A Vindication of the Stars* . . . (London, 1710).

Swift, Jonathan and Pope, Alexander, *Miscellanies in Prose and Verse* (London, 1727; 5 vols).

The True Characters of, viz., . . . *A know-all Astrological Quack* . . . (London, 1708).

Thrale [later Piozzi], Hester Lynch, *Thraliana: The Diary of Mrs Hester Lynch Thrale, 1776–1809*, ed. Katherine C. Balderston (Oxford: Oxford Unviersity Press, 1942; 2 vols).

Thrasher, William, *Jubar Astrologicum, or A True Astrological Guide* (London, 1671).

Tomkis, Thomas, *Albumazar: A Comedy*, ed. Hugh H. Dick (Berkeley, University of California Press, 1944; first published 1615).

Tonge, Ezerel [or Israel], *The Northern Star: The British Monarchy* . . . (London, 1680).

Tozer, William, *Scriptural and Hieroglyphic Observations which were foretold* . . . *by Francis Moore* . . . (London, [1812]).

Turner, Richard, *A View of the Heavens* . . ., 2nd edn (London, 1783).

Turner, Robert, *An Astrological Catechism* . . . (London, n.d.).

Tusser, Thomas, . . . *1557 Floruit: His Good Points of Husbandry*, ed. Dorothy Hartley (London: Country Life, 1931; editions throughout the sixteenth and seventeenth centuries, also in 1710 and 1744 (as *Tusser Redivivus*) and 1812).

Vicars, John, *Against William Li-lie (alias) Lillie* (London, 1652).

Walton, Richard ['the Conjurer'], *The Genuine Life, Confession, and Dying Speech of Richard Walton* . . . (Birmingham, 1733).

Ward, G. R. M. (Ed.), *Oxford University Statutes* (Oxford: Oxford University Press, 1845–51; 2 vols).

Ward, Seth, *Vindicae Academiarum* . . ., preface by John Wilkins (London, 1654).

Warren, Hardick, *Magick and Astrology Vindicated* . . . (London, 1651).

Weaver, Edmund, in *The British Telescope* . . ., an ephemeris, published annually 1723–49.

Webster, John, *Academiarum Examen* . . . (London, 1654).

——, *The Displaying of Supposed Witchcraft* . . . . (London, 1677).

Weigel, Valentin, *Astrologie Theologized* . . . (London, 1649).

Wharton, George, *Bellum Hybernicale, or Ireland's War* . . . (London, 1647).

——, in *No Merline, nor Mercurie* . . ., an almanac, published annually 1647–8; *Hemeroscopeion* . . ., 1649–54; *Hemerologium* . . . 1656–1660.

——, *The Works of that late most Excellent most Excellent Philosopher and Astronomer George Wharton*, ed. John Gadbury (London, 1683).

Whiston, William, *A New Theory of the Earth* . . . (London, 1696).

——, *A Vindication of the New Theory of the Earth* . . . (London, 1698).

——, *Astronomical Principles of Religion, Natural and Revealed* . . . (London, 1717).

White, Robert, in . . . *The Celestial Atlas* . . . , an ephemeris, published annually 1750–73, and posthumously 1774–1848.

Williams, Mrs., Astrologer . . . (London, 1790?).

Wilson, John, *The Cheats* [a play] (London, 1662).

Wing, Vincent, Letter to Lilly, 28 July 1650 (Ash. MS 423 II, 174).

——, *Harmonicon Coeleste: or the Coelestiall Harmony of the Visible World* . . . (London, 1651).

——, *An Ephemeris of the Coelestiall Motions for 7 Years, 1652–58* (London, 1652).

——, *Astronomia Instaurata* . . . (London, 1656).

——, *Astronomia Britannica* . . . (London, 1669).

——, in *[Olympia Domata]*, or, *An Almanac* . . ., published annually 1651–68, and posthumously up to 1800.

Wing, Vincent, and Leybourne, William, *Urania Practica: or, Practical Astronomie* . . . (London, 1649a).

——, *Ens Fictum Shakerlaei* . . . (London, 1649b).

Winstanley, Gerrard, *The Works of Gerrard Winstanley*, ed. George H. Sabine (Ithaca, New York: Cornell University Press, 1941).

Wood, Anthony à, *Athenae Oxonienses* . . . (London, 1691–92; 2 vols).

——, *The Life and Times of Anthony Wood* . . ., ed. Andrew Clark (Oxford: Oxford University Press, 1891–1907; 5 vols; first published in 4 vols in 1813–20).

Woodforde, James, *The Diary of a Country Parson* . . ., ed. John Beresford (London: Oxford University Press, 1924–31; 5 vols).

Worsdale, John, *Genethliacal Astrology. Comprehending an Enquiry into, and Defence of the Celestial Science* . . . (Newark, 1796).

——, *A Collection of Remarkable Nativities* . . . .*Proving the Truth and Verity of Astrology* . . . (Newark, 1799).

——, *The Nativity of Napoleon Bonaparte* . . . (Stockport, 1805).

——, *Astronomy and Elementary Philosophy* . . . (London, 1819).

——, *Celestial Philosophy, or Genethliacal Astronomy* . . . (Lincoln and London, 1828).

Worsley, Benjamin, 'Problema Physico-Astrologicum' (1657a; Hartlib Papers 26/59/1).

——, 'Physico-Astrologicall Letter' (1657b; Hartlib Papers 42/1/9).

Wren, Christopher, *Parentalia, or memoirs* . . . *chiefly of Sir Christopher Wren* . . . (London, 1750).

Wright, J., 'On the Death of the Reverend Dr John Goad' (Wood MS. 429 (47)).

[Yalden, Thomas], *Squire Bickerstaff Detected* . . . (London, 1709).

Younge [or Yonge], James, *Sidrophel Vapulans: or, The Quack Astrologer Toss'd in a Blanket* . . . (London, 1699).

——, *The Journal of James Yonge (1647–1721), Plymouth Surgeon*, ed. F. N. L. Poynter (London, 1963).

Secondary Sources

Abrams, Philip, 'History, Sociology, Historical Sociology', *Past and Present* 87 (1980) 3–16.

——, *Historical Sociology* (Shepton Mallet: Open Books, 1982).

Adorno, Theodor, 'The Stars Down to Earth: The L.A. Times Astrology Column', *Telos* 19 (1974) 13–90.

Alexander, H. G., *Religion in England 1558–1662* (London: University of London Press, 1968).

Allen, D. C., *The Star-Crossed Renaissance: The Quarrel About Astrology and its Influence in England* (New York: Octagon Books, 1973; first published 1944).

Allen, W. O. B. and McClure, Edmund, *Two Hundred Years: The History of the Society for Promoting Christian Knowledge 1698–1898* (London: 1898).

Alston, Robin C. (Ed.), *The Eighteenth-Century Short-Title Catalogue* (London: British Library, from 1983).

Apt, A. J., 'The Reception of Kepler's Astronomy in England, 1609–1650', Oxford University D.Phil. (1982).

Ariès, Phillipe, 'L'Histoire des Mentalités', pp. 402–22 in Le Goff et al. (1978).

Armytage, W. H. G., *Heavens Below. Utopian Experiments in England, 1560–1660* (London: Routledge & Kegan Paul, 1961).

Ashplant, Timothy and Wilson, Adrian, 'Present-Centred History and the Problem of Historical Knowledge', *Historical Journal* 31:2 (1988) (forthcoming).

Bailey, Francis, *An Account of the Reverend John Flamsteed, the First Astronomer-Royal; compiled from his own MSS* . . . (London, 1835).

Baker, Keith Michael, 'On the Problem of the Ideological Origins of the French Revolution', pp. 197–219 in La Capra and Kaplan (Eds) (1982).

Baker, George, *The History and Antiquities of the County of Northampton-shire* (London, 1822–30; 2 vols).

Balleine, George R., *Past Finding Out: The Tragic Story of Joanna Southcott and Her Successors* (New York: Macmillan, 1956).

Barber, Giles and Fabian, Bernard (Eds), *Buch und Buchhandel in Europa im achtzehnten Jahrhundert* (Hamburg: Ernst Hausewell, 1981).

Barclay, Oliva, 'Nicholas Culpeper: Herbalist and Astrologer', *Astrological Journal* 27 (1985) 235–7.

Barker, Felix, and Jackson, Peter, *London: Two Thousand Years of a City and its People* (London: Cassell, 1974).

Barnes, Barry, *Scientific Knowledge and Sociological Theory* (London: Routledge & Kegan Paul, 1974).

——, *Interests and the Growth of Scientific Knowledge* (London: Routledge & Kegan Paul, 1977).

Barnes, Barry and Bloor, David, 'Relativism, Rationalism and the Sociology of Knowledge', pp. 21–47 in Hollis and Lukes (Eds) (1982).

Barnes, J., Brunschweig, J., Burnyeat, M., and Schofield, M. (Eds), *Science and Speculation: Studies in Hellenistic Theory and Practice* (Cambridge: Cambridge University Press, 1982).

Barnes, Barry and Shapin, Steven (Eds), *Natural Order: Historical Studies of Scientific Culture* (London: Sage, 1979).

Barrow, Logie, *Independent Spirits: Spiritualism and English Plebeians, 1850–1910* (London: Routledge & Kegan Paul, 1986).

Barry, Jonathan, 'The Cultural Life of Bristol, 1640–1775', Oxford University D.Phil. (1986).

Bates, T. R., 'Gramsci and the Theory of Hegemony', *Journal of the History of Ideas* 36 (1975) 351–66.

Bechler, Zev (Ed.), *Contemporary Newtonian Research* (Dordrecht: D. Reidel, 1983).

Beer, A. and P. (Eds), *Kepler: Four Hundred Years* (New York: Pergamon, 1975).

Belanger, Terry, 'Publishers and Writers in Eighteenth Century England', pp. 5–25 in Rivers (Ed.) (1982).

Bennett, Tony, Mercer, Colin, and Woolacott, Jane (Eds), *Popular Culture and Social Relations* (Milton Keynes: Open University, 1986).

Berman, Morris, '"Hegemony" and the Amateur Tradition in British Science', *Journal of Social History 8 (1975) 30–50.*

——, *The Re-enchantment of the World* (Ithaca: Cornell University Press, 1981).

Bialas, V., 'Ephemerides of the Early Seventeenth Century', Vistas in Astronomy 22 (1978) 21–6.

Bigsby, C. W. E. (Ed.), *Approaches to Popular Culture* (London: Edward Arnold, 1976).

Birch, Thomas, *The History of the Royal Society of London . . .* (London, 1756–7; 4 vols).

Blagden, Cyprian, 'The Distribution of Almanacs in the Second Half of the Seventeenth Century', *Studies in Bibliography* 11 (1958) 107–116.

——, *The Stationer's Company: A History 1403–1959* (London: George Allen & Unwin, 1960).

——, 'Thomas Carnan and the Almanack Monopoly', *Studies in Bibliography* 14 (1961) 23–43.

Bocock, Robert, *Hegemony* (London: Tavistock, 1986).

Bollème, Geneviève, *Les Almanacks Populaires aux XVIIe et XVIIIe Siècles, Essai d'Histoire Sociale* (Paris: Mouton, 1969).

Bonelli, M. L. R., and Shea, W. R. (Eds), *Reasons, Experiment and Mysticism in the Scientific Revolution* (New York: Science History Publications, 1975).

Bonfield, Lloyd, Smith, Richard M., and Wrightson, Keith, *The World We Have Gained: Histories of Population and Social Structure* (Oxford: Basil Blackwell, 1986).

Bosanquet, Eustace F., *English Printed Almanacks and Prognostications: A Bibliographical History to the Year 1600* (London: Bibliographical Society, 1917).

——, 'English Seventeenth Century Almanacks', *Library* 10 (1930) 361–97.

Bouché-Leclercq, A. *L'Astrologie Grecque* (Brussels: Leroux, 1899; reprinted in Paris in 1963).

Bourdieu, Pierre, *La Distinction. Critique Sociale du Jugement* (Paris: Editions de Minuit, 1979; published in English as *Distinction* . . . (London: Routledge & Kegan Paul, 1984)).

Bowden, Mary Ellen, 'The Scientific Revolution in Astrology: the English Reformers (1558–1686)', Yale University Ph.D. (1974).

Bowles, Geoffrey, 'The Place of Newtonian Explanation in English Popular Thought, 1687–1727', Oxford University D.Phil. (1977).

Brackenridge, J. B., 'Foreword, Notes and Analytical Outline' to Kepler's 'On the More Certain Fundamentals of Astrology . . .', transl. Mary Ann Rossi, *Proceedings of the American Philosophical Society* 123 (1979) 85–116.

Brady, D., '1666: The Year of the Beast', *Bulletin of the John Rylands University Library of Manchester* 61 (1979) 314–36.

Brann, N. C. 'The Conflict Between Reason and Magic in Seventeenth Century England: A Case Study of the Vaughan-More Debate', *Huntington Library Quarterly* 43 (1980) 103–26.

Brasewell, L., 'Popular Lunar Astrology in the Late Middle Ages', *University of Ottawa Quarterly* 48 (1978) 187–94.

Brewer, John, *Party Ideology and Popular Politics at the Accession of George III* (Cambridge: Cambridge University Press, 1976).

Buck, Peter, 'Seventeenth Century Political Arithmetic: Civil Strife and Vital Statistics', *Isis* 68 (1977) 67–84.

Burghière, André, 'The Fate of the History of Mentalités in the Annales', *Comparative Studies in Society and History* 24:3 (1982) 424–37.

Burke, John G., *The Uses of Science in the Age of Newton* (Berkeley, CA: University of California Press, 1983).

Burke, Peter, 'Oblique Approaches to the History of Popular Culture', pp. 69–84 in Bigsby (Ed.) (1976).

——, 'Popular Culture in Seventeenth Century London', *London Journal* 3 (1977) 143–62.

——, *Popular Culture in Early Modern Europe* (London: Temple Smith, 1978).

——, 'Reflections on the Historical Revolution in France: The *Annales* School and British Social History', *Review* 1 (1978a) 147–64.

——, *Sociology and History* (London: George Allen & Unwin, 1980).

——, 'The "Discovery" of Popular Culture', pp. 216–26 in Samuel (Ed.) (1981).

——, 'People's History or Total History', pp. 4–8 in Samuel (Ed.) (1981a).

——, 'A Question of Acculturation?', pp. 197–204 in Scienza, *Credenze Occulte, Livelli di Cultura* (1982).

——, 'Measure of Mentalities', *Times Literary Supplement* (8 July 1983) 723.

——, 'Strengths and Weaknesses of the History of Mentalities', *History of European Ideas* 7:5 (1986) 439–51.

——, 'Revolution in Popular Culture', pp. 206–25 in Porter and Teich (Eds) (1987).

Burke, Peter, and Porter, Roy (Eds), *The Social History of Language* (Cambridge: Cambridge University Press, 1987).

Burtt, Edwin Arthur, *The Metaphysical Foundations of Physical Science: A Historical and Critical Essay* (London: Routledge & Kegan Paul, 1924).

Butler, Jon, 'Magic, Astrology and the Early American Religious Heritage, 1600–1760', *American Historical Review* 84 (1979) 317–46.

Butterfield, Herbert, *The Whig Interpretation of History* (Harmondsworth: Penguin, 1931).

Bynum, W. F., Browne, E. J., and Porter, Roy, (Eds), *Dictionary of the History of Science* (London: Macmillan, 1981).

Camden, Carroll, 'Elizabethan Almanacs and Prognostications', *Library* 12 (1931) 83–108, 194–207.

Campion, Nicholas, 'Astrology in England Before the Normans', *Astrology* 56 (1982) 51–58.

Capp, Bernard, *Astrology and the Popular Press: English Almanacs 1500–1800* (London: Faber, 1979).

——, 'The Status and Role of Astrology in Seventeenth Century England: The Evidence of the Almanac', pp. 279–90 in *Scienza, Credenze Occulte, Livelli di Cultura* (1982).

Carey, Hilary, 'Astrology and Divination in Later Medieval England', Oxford University D.Phil. (1984).

——, 'Astrology at the English Court in the Later Middle Ages', pp. 41–56 in Curry (Ed.) (1987).

Carswell, John, *From Revolution to Revolution: England 1688–1776* (London: Routledge & Kegan Paul, 1973).

Chambers, J. D., *Population, Economy and Society in Pre-Industrial Society*, ed. W. A. Armstrong (Oxford: Oxford University Press, 1972).

Chartier, Roger, 'Intellectual History or Sociocultural History? The French Trajectory', pp. 13–46 in La Capra and Kaplan (Eds) (1982; also published as 'Histoire Intellectuale et Histoire des Mentalités. Trajectoires et Questions', *Revue de Synthèse* 111–112 (1982) 277–308).

——, 'Culture as Appropriation: Popular Cultural Uses in Early Modern France', pp. 229–54 in Kaplan (Ed.) (1984).

——, *Cultural History: Between Practices and Representations* (Cambridge: Polity Press, 1988).

Christianson, John, 'Tycho Brahe's Cosmology from the *Astrologia* of 1591', *Isis* 59 (1968) 312–18.

Clark, J. C. D., *English Society 1688–1832* (Cambridge: Cambridge University Press, 1985).

——, *Revolution and Rebellion: State and Society in England in the Seventeenth and Eighteenth Centuries* (Cambridge: Cambridge University Press, 1986).

Clark, Stuart, 'Inversion, Misrule and the Meaning of Witchcraft', *Past and Present* 86 (1980) 54–97.

——, 'French Historians and Early Modern Popular Culture', *Past and Present* 100 (1983) 62–99.

——, 'The Scientific Status of Demonology', pp. 351–74 in Vickers (Ed.) (1984).

——, 'The *Annales* Historians', pp. 179–211 in Skinner (Ed.) (1985).

Clarke, Angus, 'Late Renaissance Astrology and Giovanni Antonio Magini (1555–1617)', University of London Ph.D. (1985).

——, 'Metoposcopy: An Art to Find the Mind's Construction in the Forehead', pp. 171–97 in Curry (Ed.) (1987).

Clarkson, L. A., *The Pre-Industrial Economy in England 1500–1750* (London: Batsford, 1971).

Cohen, E. H., and Ross, J. S., 'The Commonplace Book of Edmond Halley', *Notes and Records of the Royal Society of London* 40 (1985) 1–40.

Cohen, G. A., *Karl Marx's Theory of History: A Defence* (Oxford: Oxford University Press, 1978).

Coleman, D. C., *The Economy of England 1450–1750* (Oxford: Oxford University Press, 1977).

Colley, Linda, *In Defiance of Oligarchy: The Tory Party 1714–60* (Cambridge: Cambridge University Press, 1982).

——, 'The Multiple Elites of Eighteenth Century Britain', *Comparative Studies in Society and History* 29:2 (1987) 408–13.

Collins, Harry M., 'The Place of the "Core-Set" in Modern Science: Social

Contingency with Methodological Propriety in Science', *History of Science* 19 (1981) 6–19.

Collison, R., *Encyclopedias: Their History through the Ages* (New York: Hafner, 1964).

Colson, F. A., *The Days of the Week* (Cambridge: Cambridge University Press, 1926).

Cooter, Roger, 'Phrenology: the Provocation of Progress', *History of Science* 14 (1976) 211–34.

——, 'Deploying "Pseudoscience": Then and Now', pp. 237–72 in Hanen et al. (Eds) (1980).

——, *Cultural Meaning and Popular Science: Phrenology and the Organization of Consent in Nineteenth Century Britain* (Cambridge: Cambridge University Press, 1984).

——, 'The History of Mesmerism in Britain: Poverty and Promise', pp. 152–62 in Schott (Ed.) (1985).

Cornelius, Geoffrey, 'The Moment of Astrology', *Astrology* 57 (1983–84) 97–112, 140–51; 58 (1984) 14–24, 85–95; 59 (1985) 42–9, 207–21.

——, 'A Modern Astrological Perspective', pp. 864–71 in Lilly (1985).

Corrigan, Philip and Sayer, Derek, *The Great Arch: English State Formation as Cultural Revolution* (Oxford: Basil Blackwell, 1985).

Corsi, Pietro and Weindling, Paul (Eds), *Information Sources in the History of Science and Medicine* (London: Butterworths, 1983).

Couttie, Bob, *Forbidden Knowledge* (Cambridge: Lutterworth, 1988).

Cowling, T. G., *Isaac Newton and Astrology* (Leeds: University of Leeds Press, 1977).

——, 'Astrology, Religion and Science', *Quarterly Journal of the Royal Astronomical Society* 23 (1982) 515–26.

Cressy, David, *Literacy and the Social Order: Reading and Writing in Tudor and Stuart England* (Cambridge: Cambridge University Press, 1980).

Croix, Geoffrey de Ste, *The Class Struggle in the Ancient Greek World* (London: Duckworth, 1981).

——, 'Class in Marx's Conception of History, Ancient and Modern', *New Left Review* 146 (1984) 94–111.

Crosland, M. (Ed.), *The Emergence of Science in Western Europe* (London: Macmillan, 1975).

Cumont, Franz, *Astrology and Religion Among the Greeks and Romans* (New York: Dover, 1960; first published 1912).

Curry, Patrick, 'Revisions of Science of Magic', *History of Science* 23 (1985) 299–325.

——, 'Afterword' and 'Bibliographical Appendix', pp. 856–64 in Lilly (1985a).

——, 'The Decline of Astrology in Early Modern England, 1642–1800', University of London Ph.D. ([completed] 1986).

—— (Ed.), *Astrology, Science and Society: Historical Essays* (Woodbridge, Suffolk: 1987); including his 'Saving Astrology in Restoration England: "Whig" and "Tory" Reforms', pp. 245–60.

——, 'The Astrologers' Feasts', *History Today* 38 (April 1988) 17–22.

Darby, H. C., *The Changing Fenland* (Cambridge: Cambridge University Press, 1983).

Darnton, Robert, *Mesmerism and the End of the Enlightenment in France* (Cambridge, MA: Harvard University, 1968).

——, *The Great Cat Massacre and Other Episodes in French Cultural History* (Harmondsworth: Penguin, 1985; first published 1984).

——, 'The Symbolic Element in History', *Journal of Modern History* 58:1 (1986) 218–34.

Davis, Natalie Zemon, *Society and Culture in Early Modern France* (Stanford, CA: Stanford University Press, 1975).

Dawson, W. Harbutt, 'An Old Yorkshire Astrologer and Magician, 1694–1760', *Reliquary* 23 (1882–3) 197–202.

Debus, Allen G., 'Alchemy and the Historian of Science', *History of Science* 6 (1967) 128–37.

——, *Science and Education in the Seventeenth Century* (London: Macdonald, 1970).

—— (Ed.), *Science, Medicine and Society in the Renaissance: Essays in Honour of Walter Pagel* (London: Heinemann, 1972; 2 vols).

——, 'Scientific Truth and Occult Tradition: The Medical World of Ebenezer Sibly', *Medical History* 26 (1982) 259–78.

Dewhurst, Kenneth, *John Locke (1632–1704): Physician and Philosopher: A Medical Biography* . . . (London: Wellcome Library, 1963).

Dick, H. G., 'Students of Physick and Astrology: A Survey of Astrological Medicine in the Age of Science', *Journal of the History of Medicine* 1 (1946) 300–15, 419–33.

Dickson, David, 'Science and Political Hegemony in the Seventeenth Century', *Radical Science Journal* 8 (1979) 7–37.

Dijksterhuis, E. J., *The Mechanization of the World Picture* (Oxford: Clarendon Press, 1961).

Dobbs, B. J. T., *The Foundations of Newton's Alchemy* (Cambridge: Cambridge University Press, 1975).

Donnelly, F. K., 'Ideology and Early English Working Class History: Edward Thompson and his Critics', *Social History* 2 (1976) 219–38.

Dreyfus, Hubert L. and Rabinow, Paul, *Michel Foucault: Beyond Structuralism and Hermeneutics* (Hassocks: Harvester Press, 1982).

Duboscq, Guy, Plongeron, Bernard, and Robert, Daniel (Eds), *La Réligion Populaire* (Paris, 1980; Colloques Internationale du CNRS, No. 576).

Duby, Georges, 'Ideologies in Social History', pp. 151–65 in Le Goff and Nora (Eds) (1985).

Eade, J. C., *The Forgotten Sky: A Guide to Astrology in English Literature* (Oxford: Clarendon Press, 1984).

Eagles, Christina M., 'The Mathematical Work of David Gregory, 1659–1708', University of Edinburgh Ph.D. (1977).

Eagleton, Terry, *The Function of Criticism* (London: Verso, 1984).

Easlea, Brian, *Witch-hunting, Magic and the New Philosophy: An Introduction to Debates of the Scientific Revolution, 1450–1750* (Brighton: Harvester Press, 1980).

Eddy, W. A., 'The Wits vs. John Partridge, Astrologer', *Studies in Philology* 29 (1932) 29–40.

Elmer, Peter, 'The Library of Dr John Webster: The Making of a Seventeenth Century Radical', *Medical History*, Supplement No. 6 (1986).

Englefield, W. A. D., *The History of the Painters-Stainers Company of London* (London: Hazell, Watson & Viney, 1923).

Entered at Stationer's Hall. A Sketch of the History and Privileges of The Company of Stationers . . . (London, 1871).

'Espinasse, Margaret, *Robert Hooke* (London: Heinemann, 1956).

Evans, Edward, *Historical and Bibliographical Account of Almanacks, Directories &c. &c. Published in Ireland From the Sixteenth Century* (Blackrock: Carraig Books, 1976; first published 1897).

Febvre, Lucien, *The Problem of Unbelief in the Sixteenth Century: The Religion of Rabelais* (Cambridge, MA: Harvard University Press, 1983; first published as *Le Problème de l'Incroyance au XVIe Siècle. La Réligion de Rabelais* (Paris, 1952, 1968)).

Feingold, Mordechai, *The Mathematicians' Apprenticeship. Science, Universities and Society in England, 1560–1640* (Cambridge: Cambridge University Press, 1984).

Feisenberger, H. A. (Ed.), *Sales Catalogues of Libraries of Eminent Persons*, Vol. 2: *Scientists* (London: Mansell, 1975; 12 vols, 1971–5).

Festugière, Le R. P., *La Révélation d'Hermès Trismégiste*, Vol. 1: *L'Astrologie et les Sciences Occultes* (Paris: Librarie Lecoffre, 1950; 4 vols, 1950–4).

Feyerabend, Paul, *Science in a Free Society* (London: New Left Books, 1978); including 'The Strange Case of Astrology' (pp. 91–5) and 'The Spectre of Relativism' (pp. 79–86).

——, *Farewell to Reason* (London: Verso, 1987).

Field, J. V., 'A Lutheran Astrologer: Johannes Kepler', *Archive for History of Exact Sciences* 31 (1984) 189–272.

——, 'Astrology in Kepler's Cosmology', pp. 143–70 in Curry (Ed.) (1987).

——, *Kepler's Geometric Cosmology* (London: Athlone, 1987a).

Fish, Stanley, *Is There a Text in this Class?* (Cambridge, MA: Harvard University Press, 1980).

Fletcher, Anthony and Stevenson, John (Eds), *Order and Disorder in Early Modern England* (Cambridge: Cambridge University Press, 1985).

Force, James E., *William Whiston: Honest Newtonian* (Cambridge: Cambridge University Press, 1985).

Fowles, John, *A Maggot* (London: Jonathan Cape, 1985).

Gardner, F. Leigh, *A Catalogue Raisonné of Works on the Occult Sciences* (London: privately printed, 1911).

Garin, Eugenio, *Astrology in the Renaissance: The Zodiac of Life* (London: Routledge & Kegan Paul, 1983; first published as *Lo Zodiaco della Vita. La Polemica sull' astrologia dal Trecento al Cinquecento* (Rome, 1976).

Garrett, Clarke, *Respectable Folly: Millenarians and the French Revolution in France and England* (Baltimore: Johns Hopkins University Press, 1975).

——, 'Swedenborg and the Mystical Enlightenment in Late Eighteenth Century England', *Journal of the History of Ideas* 45 (1984) 67–81.

Gascoigne, John, 'Politics, Patronage and Newtonianism: The Cambridge Experience', *Historical Journal* 27 (1984) 1–24.

——, 'Anglican Latitudinarianism and Political Radicalism in the Late Eighteenth Century', *History* 71 (1986) 22–38.

Gauquelin, Michel, *The Truth about Astrology* (Oxford: Basil Blackwell, 1984).

Geertz, Hildred, 'An Anthropology of Religion and Magic, I', *Journal of Interdisciplinary History* 6 (1975) 71–89.

Geneva, Ann, 'Diary', *London Review of Books* (26 November 1987) 25.

——, 'England's Propheticall Merline Decoded: A Study of the Symbolic Art of Astrology in Seventeenth Century England', State University of New York (Stony Brook) Ph.D. (1988).

Genuth, Sara Schechner, 'Comets, Teleology and the Relation of Chemistry to Cosmology in Newton's Thought', *Annali Dell' Istituto e Museo di Storia della Scienza di Firenze* 10:2 (1985) 131–65.

Gibson, Strickland (Ed.), *Statuta Antiqua Universitatis Oxoniensis* (Oxford: Clarendon Press, 1931).

Gillispie, Charles C. (Ed.), *Dictionary of Scientific Biography* (New York: Charles Scribner's Sons, 1974; 16 vols).

Ginzburg, Carlo, *The Cheese and the Worms: The Cosmos of a Sixteenth Century Miller* (London: Routledge & Kegan Paul, 1980; first published as *Il formaggio e i vermi: Il cosmo di un mugnaio del '500* (Italy, 1976)).

Gismondi, Michael A., '"The Gift of Theory": A Critique of the *histoire des mentalités*', *Social History* 10 (1985) 211–30.

Golby, J. M. and Purdue, A. W., *The Civilization of the Crowd: Popular Culture in England, 1750–1900* (London: Batsford, 1984).

Gooding, David, Pinch, Trevor, and Schaffer, Simon (Eds), *The Uses of Experiment: Studies in the Natural Sciences* (Cambridge: Cambridge University Press, 1988).

Gouk, Penelope, 'Music and Natural Philosophy in the Early Royal Society', University of London Ph.D. (1982).

Graham, Walter, *English Literary Periodicals* (New York: Thomas Nelson, 1930).

Gramsci, Antonio, *Selections from the Prison Notebooks* (London: Lawrence & Wishart, 1971).

Graubard, Mark, 'Astrology's Demise and Its Bearing on the Decline and Death of Beliefs', *Osiris* 13 (1958) 210–61.

——, *Witchcraft and the Nature of Man* (Lanham: University Press of America, 1985).

Gray, Robert Q., *The Labour Aristocracy in Victorian Edinburgh* (Oxford: Clarendon Press, 1976).

Greaves, Richard L. and Zaller, Robert (Eds), *Biographical Dictionary of British Radicals in the Seventeenth Century* (Brighton: Harvester Press, 1982–4; 3 vols).

Greg, W. W., *Some Aspects and Problems of London Publishing between 1550 and 1650* (Oxford: Clarendon Press, 1956).

——, *Licensers for the Press, &c. to 1640: A Biographical Index Based Mainly on Arber's Transcript of the Registers of the Company of Stationers* (Oxford: Oxford Bibliographical Society, 1962).

Greyerz, Kaspar von (Ed.), *Religion and Society in Early Modern Europe 1500–1800* (London: George Allen & Unwin, 1984).

Growell, A., *Three Centuries of English Booktrade Bibliography* (New York: Dibdin Club, 1903).

Gunther, R. W. T., *Early Science in Oxford*, Vol. 12: *Dr Plot and the Correspondence of the Philosophical Society of Oxford*, and Vol. 4: *The Philosophical Society* (Oxford: for the Subscribers, 1923–45; 14 vols).

Hacking, Ian, *The Emergence of Probability* (Cambridge: Cambridge University Press, 1975).

Halbronn, Jacques, 'The Revealing Process of Translation and Criticism in the History of Astrology', pp. 197–218 in Curry (Ed.) (1987).

——, 'Introduction Bibliographique à l'Etude de l'Astrologie Française', Université de Paris 9e Thèse d'Etat (1989).

Halbronn, Jacques and Hutin, Serge, *Histoire de l'Astrologie* (Paris: Artefact, 1986).

Hall, A. R., *The Revolution in Science, 1500–1750* (London: Collins, 1983; rev. edn. of *The Scientific Revolution, 1500–1800* (1954)).

——, 'On Whiggism', *History of Science* 21 (1983) 45–59.

Hall, Stuart, 'Notes on Deconstructing "the Popular"', pp. 225–40 in Samuel (Ed.) (1981).

——, 'Popular Culture and the State', pp. 22–49 in Bennett et al. (Eds) (1986).

Hanen, Marsha P., Osler, Margaret J., and Weyant, Robert G. (Eds), *Science, Pseudo-science and Society* (Waterloo, Ontario: Wilfred Laurier University, 1980).

Hansen, Bert, 'Science and Magic', pp. 483–506 in Lindberg (Ed.) (1978).

Harris, Michael, *London Newspapers in the Age of Walpole: A Study of the Origins of the Modern English Press* (Madison, WI: Fairleigh Dickinson University, 1987).

Harrison, J. F. C., *The Second Coming: Popular Millenarianism 1780–1850* (London: Routledge & Kegan Paul, 1979).

Henry, John, 'Occult Qualities and the Experimental Philosophy: Active Principles in Pre-Newtonian Matter Theory', *History of Science* 24 (1986) 335–81.

——, 'Henry More *vs.* Robert Boyle: The Spirit of Nature and the Nature of Providence', forthcoming in S. Hutton (Ed.) (1989).

Hesse, Mary, *Revolutions and Reconstructions in the Philosophy of Science* (Brighton: Harvester Press, 1980).

Heywood, Abel, *Three Papers on English Printed Almanacks* (?London: privately printed, 1904).

Hilaire, Yves-Marie, *La Réligion Populaire: Aspects du Christianisme Populaire à Travers L'Histoire* (Lille: Université de Lille, 1981).

Hill, Christopher, *Intellectual Origins of the English Revolution* (Oxford: Clarendon Press: 1965).

——, *The Century of Revolution, 1603–1714* (London: Abacus, 1974; first published 1961).

——, *The World Turned Upside Down: Radical Ideas during the English Revolution* (Harmondsworth: Penguin, 1975; first published 1972).

——, *Some Intellectual Consequences of the English Revolution* (London: Weidenfeld & Nicolson, 1980).

——, 'Science and Magic in Seventeenth Century England', pp. 176–93 in Samuel and Jones (Eds) (1982).

——, *The Experience of Defeat: Milton and Some Contemporaries* (London: Faber, 1984).

——, *The Collected Essays of Christopher Hill*, Vol. 1: *Writings and Revolution in Seventeenth Century England* (Brighton: Harvester Press, 1985).

Hirst, Desirée, *Hidden Riches: Traditional Symbolism from the Renaissance to Blake* (London: Eyre & Spottiswoode, 1964).

Hirst, Paul, 'Witchcraft Today and Yesterday', *Economy and Society* 11:4 (1982) 428–48.

216

——, 'Is it Rational to Reject Relativism?' pp. 85–103 in Overing (Ed.) (1985).

——, *Marxism and Historical Writing* (London: Routledge & Kegan Paul, 1985a).

Hirst, Paul and Wooley, Penny, *Social Relations and Human Attributes* (London: Tavistock, 1982).

Hobhouse, Stephen, (Ed.), *Select Mystical Writings of William Law* (London: Rockliff, 1948).

Hobsbawm, Eric J., 'The Revival of Narrative: Some Comments', *Past and Present* 86 (1980) 3–8.

Hobsbawm, Eric J. and Ranger, Terence (Eds), *The Invention of Tradition* (Cambridge: Cambridge University Press, 1983).

Hole, Christina, *English Folklore* (London: Batsford, 1940).

Holland, A.J. (Ed.), *Philosophy, Its History and Historiography* (Dordrecht: D. Reidel, 1985).

Hollis, Martin and Lukes, Steven (Eds), *Rationality and Relativism* (Oxford: Basil Blackwell, 1982).

Holmes, Clive, 'Popular Culture? Witches, Magistrates, and Divines in Early Modern England', pp. 85–111 in Kaplan (Ed.) (1984).

——, 'Drainers and Fenmen', pp. 116–95 in Fletcher and Stevenson (Eds) (1985).

Holmes, Geoffrey, 'Science, Reason and Religion in the Age of Newton', *British Journal for the History of Science* 11 (1978) 164–71.

——, *Augustan England: Professions, State and Society, 1680–1730* (London: George Allen & Unwin, 1982).

Hoppen, K. Theodore, *The Common Scientist in the Seventeenth Century: A Study of the Dublin Philosophical Society* (London: Routledge & Kegan Paul, 1970).

——, 'The Nature of the Early Royal Society', *British Journal for the History of Science* 9 (1976) 1–24, 243–73.

Horn, Pamela, *The Rural World 1780–1850: Social Change in the English Countryside* (London: Hutchinson, 1980).

Hoskin, Michael, 'Stukeley's Cosmology and the Newtonian Origins of Olber's Paradox', *Journal for the History of Astronomy* 16 (1985) 77–112.

Howe, Ellic, 'The Stationers' Company Almanacks: A Late Eighteenth Century Printing and Publishing Operation', pp. 195–209 in Barber and Fabian (Eds) (1981).

——, *Astrology and the Third Reich* (Wellingborough: Aquarian, 1984; first published as *Urania's Children: The Strange World of the Astrologers* (London, 1967)).

Hultin, N.C., 'Medicine and Magic in the Eighteenth Century: The Diaries of John Woodforde', *Journal of the History of Medicine* 30 (1975) 349–66.

Hunter, Michael, *John Aubrey and the Realm of Learning* (London: Duckworth, 1975).

——, *Science and Society in Restoration England* (Cambridge: Cambridge University Press, 1981).

——, *The Royal Society and its Fellows 1660–1700: The Morphology of an Early Scientific Organization* (Chalfont St Giles: BSHS, 1982).

——, *Elias Ashmole, 1617–1692: The Founder of the Ashmolean Museum and his World: A Tercentenary Exhibition* (Oxford: Ashmolean Museum, 1983).

——, 'Science and Astrology in Seventeenth Century England: An Unpublished Polemic by John Flamsteed', pp. 261–300 in Curry (Ed.) (1987).

Hunter, Michael and Gregory, Annabel, *An Astrological Diary of the Seventeenth Century: Samuel Jeake of Rye* (1652–1699) (Oxford: Clarendon Press, 1988).

Hutchison, Keith, 'What Happened to Occult Qualities in the Scientific Revolution?' *Isis* 73 (1982) 233–53.

——, 'Supernaturalism and the Mechanical Philosophy', *History of Science* 21 (1983) 297–333.

——, 'Towards a Political Iconology of the Copernican Revolution', pp. 95–142 in Curry (Ed.) (1987).

Hutin, Serge, *Les Disciples Anglais de Jacob Boehme aux XVIIe et XVIIIe Siècles* (Paris: Denoel, 1960).

——, *L'Histoire de l'Astrologie, science ou superstition* (Verviers: Marabout Université, 1970).

Hutton, Patrick H., 'The History of Mentalities: The New Map of Cultural History', *History and Theory* 20 (1981) 237–59.

Hutton, Sarah, (Ed.), *Mysticism and Mechanism: Tercentenary Studies of Henry More (1614–1687)* (The Hague: Martinus Nijhoff, 1989).

Iggers, Georg, *New Directions in European Historiography* (Middletown, CT: Wesleyan University, 1974).

Ingram, Martin, 'Ridings, Rough Music and the "Reform of Popular Culture" in Early Modern England', *Past and Present* 105 (1984) 79–113.

Inkster, Ian, 'Science and Society in the Metropolis: A Preliminary Examination of the Social and Institutional Context of the Askesian Society of London, 1796–1807', *Annals of Science* 34 (1977) 1–32.

——, 'Introduction: Aspects of the History of Science and Science Culture in Britain, 1780–1850 and Beyond', pp. 11–54 in Inkster and Morrell (Eds) (1983).

Inkster, Ian, and Morrell, Jack (Eds), *Metropolis and Province: Science in British Culture, 1780–1850* (London: Hutchinson, 1983).

Ives, E. W., 'The World of the Common Man', *Times Higher Education Supplement* (8 February 1980) 16.

Jacob, James R., 'Robert Boyle and Subversive Religion in the Early Restoration', *Albion* 6 (1974) 275–93.

——, 'Boyle's Circle in the Protectorate: Revelation, Politics and the Millenium', *Journal of the History of Ideas* 38 (1977a) 131–40.

——, *Robert Boyle and the English Revolution: A Study in Social and Intellectual Change* (New York: Burt Franklin, 1977b).

——, 'Boyle's Atomism and the Restoration Assault on Pagan Naturalism', *Social Studies of Science* 8 (1978) 211–33.

——, 'Restoration Ideologies and the Royal Society', *History of Science* 18 (1980) 25–38.

——, *Henry Stubbe, Radical Protestantism and the Early Enlightenment* (Cambridge: Cambridge University Press, 1983).

Jacob, James R. and Margaret C., 'The Anglican Origins of Modern Science: The Metaphysical Foundations of the Whig Constitution', *Isis* 71 (1980) 251–67.

Jacob, Margaret C., 'John Toland and the Newtonian Ideology', *Journal of the Warburg and Courtauld Institutes* 32 (1969) 307–31.

——, *The Newtonians and the English Revolution, 1679–1720* (Hassocks: Harvester Press, 1976a).

——, 'Millenialism and Science in the Late Seventeenth Century', *Journal of the History of Ideas* 37 (1976b) 335–41.

——, 'Newtonianism and the Origins of the Enlightenment: A Reassessment', *Eighteenth Century Studies* 11 (1977) 1–25.

——, 'Newtonian Science and the Radical Enlightenment', *Vistas in Astronomy* 22 (1979) 545–55.

——, *The Radical Enlightenment: Pantheists, Freemasons and Republicans* (London: George Allen & Unwin, 1981).

——, 'Science and Social Passion: The Case of Seventeenth Century England', *Journal of the History of Ideas* 43 (1982) 331–9.

Jarrett, Derek, *England in the Age of Hogarth* (New Haven, CT: Yale University Press, 1974).

Jayawardene, S. A., *Reference Books for the Historian of Science: A Handlist* (London: Science Museum Library, 1982).

Jeaffreson, J. C., (Ed.), *Middlesex County Records* (London, 1887–1902; 4 vols).

Jenks, Stuart, 'Astrometeorology in Middle Ages', *Isis* 74 (1983) 185–210.

Jevons, F. R., 'Paracelsus' Two-Way Astrology', *British Journal for the History of Science* 2 (1964) 139–55.

Jobe, Thomas Harmon, 'The Devil in Restoration Science: The Glanvill-Webster Witchcraft Debate', *Isis* 72 (1981) 343–56.

Johnson, Richard, McLennan, Gregor, Schwartz, Bill, and Sutton, David, *Making Histories: Studies in History-writing and Politics* (London: Hutchinson, 1982).

Joll, James, *Gramsci* (London: Fontana, 1977).

Jones, Gareth Stedman, 'From Historical Sociology to Theoretical History', *British Journal of Sociology* 27 (1976) 295–305.

——, *Languages of Class: Studies in English Working Class History* (Cambridge: Cambridge University Press, 1983).

Jones, Marc Paul, 'Henry Andrews' (unpublished paper, 1987).

Jones, R. F., *Ancients and Moderns: A Study of the Rise of the Scientific Movement in Seventeenth Century England*, 2nd edn (Gloucester, MA: Peter Smith, 1961).

Jones, Roger S., *Country and Court: England 1658–1714* (London: Edward Arnold, 1978).

——, *Physics as Metaphor* (London: Abacus, 1983).

Kaplan, Steven L. (Ed.), *Understanding Popular Culture: Europe from the Middle Ages to the Nineteenth Century* (Amsterdam: Mouton, 1984).

Kargon, Robert H., 'John Graunt, Francis Bacon and the Royal Society: The Reception of Statistics', *Journal of the History of Medicine and Allied Sciences* 18 (1963) 337–48.

Kelly, John Thomas, 'Practical Astronomy During the Seventeenth Century: A Study of Almanac-Makers in America and England', Harvard University Ph.D. (1977).

Kenyon, J. P., *Revolution Principles: The Politics of Party, 1689–1720* (Cambridge: Cambridge University Press, 1977).

Kingston, Alfred, *A History of Royston* (London: Elliott Sock, 1906).

Kitchin, George, *Sir Roger L'Estrange: A Contribution to the History of the Press in the Seventeenth Century* (New York: Augustus M. Kelley, 1971; first published in London, 1913).

Klibansky, Raymond, Panofsky, Erwin, and Saxl, Fritz, *Saturn and Melancholy. Studies in the History of Natural Philosophy, Religion and Art* (London: Nelson, 1964).

Kocher, Paul, *Science and Religion in Elizabethan England* (San Marino: Huntington Library, 1953).

Kollerstrom, Nick, Astro-Chemistry: A Study of Metal–Planet Affinities (London: Element, 1984).

Kubrin, David, 'Newton's Inside Out! Magic, Class Struggle, and the Rise of Mechanism in the West', pp. 96–121 in Woolf (Ed.) (1981).

——, '"Burning Times," Isaac Newton, and the War Against the Earth', pp. 38/1–38/34 in *Proceedings of the Conference 'Is the Earth a Living Organism?'* (Amherst: University of Massachusetts, 16 August 1985).

Kuhn, Thomas, *The Structure of Scientific Revolutions* (Princeton: Princeton University, 1962).

——, Sherman Memorial Lectures (London, 1987: private MS).

Labrousse, Elizabeth, *L'Entrée de Saturne au Lion: L'Eclipse de Soleil du 12 Aôut 1654* (The Hague: Martinus Nijhoff, 1974).

——, *Bayle* (Oxford: Oxford University Press, 1983).

La Capra, Dominick and Kaplan, Steven L. (Eds.), *Modern European Intellectual History: Reappraisals and New Perspectives* (Cornell: Cornell University Press, 1982).

Laclau, Ernesto and Mouffe, Chantal, *Hegemony and Socialist Strategy: Towards a Radical Democratic Politics* (London: Verso, 1985).

——, 'Post-Marxism Without Apologies', *New Left Review* 166 (1987) 79–106.

Larkey, Sanford V., 'Astrology and Politics in the First Years of Elizabeth's Reign', *Bulletin of the Instutute of the History of Medicine* 3 (1935) 171–86.

Larkin, Philip, *Required Writing: Miscellaneous Pieces 1955–1982* (London: Faber, 1983).

Larner, Christina, *Witchcraft and Religion: The Politics of Popular Belief* (Oxford: Basil Blackwell, 1984).

Larrain, Jorge, *The Concept of Ideology* (London: Hutchinson, 1979).

Laslett, Peter, 'A One-Class Society,' pp. 196–221 in Neale (Ed.) (1983).

Leeuwen, Henry G. van, *The Problem of Certainty in English Thought 1630–1690*: The Hague: Martinus Nijhoff, 1963).

Le Goff, Jacques, 'Mentalities: A History of Ambiguities,' pp. 116–180 in Le Goff and Nora (Eds) (1985; first published as 'Les mentalités: Une histoire ambigue', pp. 76–94 in the first edition of Le Goff and Nora (1985)).

Le Goff, Jacques, Chartier, Roger, and Revel, Jacques (Eds), *La Nouvelle Historie* (Paris: Retz, 1978).

Le Goff, Jacques and Nora, Pierre, *Constructing the Past: Essays in Historical Methodology* (Cambridge: Cambridge University Press, 1985; first published as *Faire de L'Histoire* (Paris, 1974)).

Leventhal, Herbert, *In the Shadow of the Enlightenment: Occultism and Renaissance Science in Eighteenth Century America* (New York: New York University, 1976).

Lindberg, David C. (Ed.), *Science in the Middle Ages* (Chicago: Chicago University Press, 1978).

Lipton, J. D., 'Astrology', pp. 30–1 in Bynum et al. (Eds) (1981).

Long, A. A., 'Astrology: Arguments Pro and Contra', pp. 165–92 in Barnes et al. (Eds) (1982).

Lloyd, G. E. R., *Magic, Reason and Experience: Studies in the Origins and Development of Greek Science* (Cambridge: Cambridge University Press, 1979).

MacCubbin, Robert P. (Ed.), *Unauthorized Sexual Behaviour during the Enlightenment: Eighteenth Century Life* 9 (May 1985).

MacDonald, Michael, *Mystical Bedlam: Madness, Anxiety and Healing in Seventeenth Century England* (Cambridge: Cambridge University Press, 1982).

——, 'Religion, Social Change, and Psychological Healing in England, 1600–1800', pp. 101–25 in Sheils (Ed.) (1982a).

——, 'Anthropological Perspectives on the History of Science and Medicine', pp. 61–80 in Corsi and Weindling (Eds) (1983).

McGregor, J. F. and Reay, B. (Eds), *Radical Religion in the English Revolution* (Oxford: Clarendon Press, 1984).

MacIntosh, Christopher, *The Astrologers and their Creed: An Historical Outline* (London: Hutchinson, 1969).

Mackay, Charles, *Memoirs of Extraordinary Popular Delusions* (London, 1841).

McKendrick, Neil, Brewer, John, and Plumb, J. H., *The Birth of a Consumer Society: The Commercialization of Eighteenth Century England* (London: Europa Publications, 1982).

McKeon, Michael, *Politics and Poetry in Restoration England: The Case of Dryden's 'Annus Mirabilis'* (Cambridge, MA: Harvard University Press, 1975).

McLennan, Gregor, *Marxism and the Methodologies of History* (London: Verso, 1981).

Malament, Barbara C. (Ed.), *After the Reformation: Essays in Honour of J. H. Hexter* (Manchester: Manchester University Press, 1980).

Malcolmson, Robert W., *Popular Recreations in English Society 1700–1850* (Cambridge: Cambridge University Press, 1973).

——, *Life and Labour in England, 1700–1780* (London: Hutchinson, 1981).

Marlowe, Christopher, *The Fen Country* (London: Cecil Palmer, 1925).

Marx, Karl, *The Eighteenth Brumaire of Louis Bonaparte* (London: Lawrence & Wishart, 1984).

Mason, Wilmer G., 'The Annual Output of Wing-listed Titles, 1649–1684', *Library* 29 (1974) 219–20.

Mathias, Peter G. (Ed.), *Science and Society 1600–1900* (Cambridge: Cambridge University Press, 1972).

Matthews, William (Ed.), *British Diaries. An Annotated Bibliography of British Diaries Written between 1442–1942* (Berkeley, CA: University of California, 1950).

——, *British Autobiographies. An Annotated Bibliography of British Autobiographies Published or Written Before 1951* (Berkeley, CA: University of California, 1955).

Mayhew, George P., 'The Early Life of John Partridge', *Studies in English Literature* 1 (1961) 31–42.

Medick, Hans, 'Plebeian Culture in the Transition to Capitalism', pp. 84–113 in Samuel and Jones (Eds) (1982).

Mendelsohn, Everett, Weingart, Peter, and Whitley, Richard (Eds), *The Social Production of Scientific Knowledge* (Dordrecht: D. Reidel, 1977).

Merchant, Carolyn, *The Death of Nature: Women, Ecology, and the Scientific Revolution* (San Francisco: Harper & Row, 1980).

Midelfort, H. C. Erik, 'Witchcraft, Magic and the Occult', pp. 183–209 in Ozment (Ed.) (1982).

Monter, William, *Ritual, Myth and Magic in Early Modern Europe* (Brighton: Harvester, 1983).

Montgomery, John Warwick, 'Cross, Constellation and Crucible: Lutheran Astrology and Alchemy in the Age of the Reformation', *Ambix* 11 (1963) 65–86.

Mouffe, Chantal (Ed.) *Gramsci and Marxist Theory* (London: Routledge & Kegan Paul, 1979); including her 'Hegemony and Ideology in Gramsci', pp. 168–204.

Muchembled, Robert, *Culture populaire et culture des élites dans la France moderne* (Paris: Flammarion, 1978).

Muddiman, J. G., *The King's Journalist, 1659–89: Studies in the Reign of Charles II* (London: Bodley Head, 1923).

Müller-Jahnke, Wolf-Dieter, 'Astrologische-Magische Theorie und Praxis in Der Heilkunde de Frühen Neuzeit', *Sudhoffs Archiv* 25 (1985).

Mulvey, Laura, 'Changes: Thoughts on Myth, Narrative and Historical Experience', *History Workshop Journal* 23 (1987) 3–19.

Naylor, P. I. H., *Astrology: An Historical Examination* (London: Robert Maxwell, 1967).

Neale, R. S., *Class in English History, 1680–1850* (Oxford: Basil Blackwell, 1981).

—— (Ed.), *History and Class: Essential Readings in Theory and Interpretation* (Oxford: Basil Blackwell, 1983).

Nelson, N. H., 'Astrology, *Hudibras*, and the Puritans', *Journal of the History of Ideas* 37 (1976) 521–36.

Neuberg, Victor E., *Popular Education in Eighteenth Century England* (London: Woburn, 1971).

Neugebauer, Otto, 'The Study of Wretched Subjects', *Isis* 42 (1951) 111 reprinted in Neugebauer (1983).

——, *The Exact Sciences in Antiquity* (New York: Dover, 1969; a reprint of the 2nd edn of 1957).

——, *Astronomy and History: Selected Essays* (New York and Berlin: Springer, 1983).

Newell, Philip, *Greenwich Hospital: A Royal Foundation, 1692–1983* (London: Trustees of Greenwich Hospital, 1984).

Nicolson, Marjorie Hope, 'English Almanacks and the New Astronomy', *Annals of Science* 4 (1939) 1–33.

——, *Pepys' Diary and the New Science* (Charlottesville, VA: University Press of Virginia, 1965).

North, J. D., 'Astrology and the Fortunes of Churches,' *Centaurus* 24 (1980) 181–211.

——, *Horoscopes and History* (London: Warburg Institute, 1985).

——, 'Medieval Concepts of Celestial Influence', pp. 5–18 in Curry (Ed.) (1987).

Obelkevitch, James, *Religion and Rural Society: South Lindsey, 1825–1875* (Oxford: Clarendon Press, 1976).

——, *(Ed.), Religion and the People, 800–1700* (Chapel Hill, NC: University of North Carolina, 1979).

——, 'Proverbs and Social History', pp. 43–72 in Burke and Porter (Eds) (1987).

Olson, Richard G., 'Tory-High Church Opposition to science and Scientism in the Eighteenth Century: The Works of John Arbuthnot, Jonathan Swift, and Samuel Johnson', pp. 171–204 in Burke, John G. (1983).

Opie, Iona and Peter, *The Lore and Language of Schoolchildren* (Oxford: Clarendon Press, 1959).

Overing, Joanna (Ed.), *Reason and Morality* (London: Tavistock, 1985).

Ozment, Steven (Ed.), *Reformation Europe: A Guide to Research* (St Louis, MO: Center for Reformation Research, 1982).

Pagel, Walter, 'The Vindication of Rubbish', *Middlesex Hospital Journal* 45 (1945) 42–5 reprinted in Pagel (1985).

——, *Paracelsus: An Introduction to Philosophical Medicine in the Era of the Renaissance* (Basel: John Wiley, 1982; first published in 1958).

——, *Joan Baptista Van Helmont: Reformer of Science and Medicine* (Cambridge: Cambridge University Press, 1982).

——, *Religion and Neoplatonism in Renaissance Medicine*, ed. Marianne Winder (London: Variorum Reprints, 1985).

Pagliaro, Harold E. (Ed.), *Irrationalism in the Eighteenth Century: Studies in Eighteenth Century Culture*, Vol. 2 (1972).

Parker, Derek, *Familiar to All: William Lilly and Astrology in the Seventeenth Century* (London: Jonathan Cape, 1975).

Parker, Derek and Julia, *A History of Astrology* (London: Andre Deutsche, 1983).

Parr, Johnstone, *Tamburlaine's Malady, and Other Essays on Astrology in Elizabethan Drama* (Alabama: University of Alabama, 1953).

Patrides, C. A. and Waddington, Raymond B. (Eds), *The Age of Milton* (Manchester: Manchester University Press, 1980).

Patte, Frank A., 'Mesmer's Medical Dissertation and its Debt to Mead's *De Imperio Solic ac Lunae*', *Journal of the History of Medicine and Allied Sciences* 11 (1956) 275–86.

Paulin, Tom, 'Clare in Babylon', *Times Literary Supplement* (20–26 June 1986) 675–6.

Payne, Harry C., 'Elite vs. Popular Mentality in the Eighteenth Century', *Studies in Eighteenth Century Culture* 8 (1979) 3–32.

Peacock, M., 'Folklore and Legends of Lincolnshire', (unpublished MS in the Folklore Society Library).

Perkin, Harold, *The Origins of Modern English Society 1780–1880* (London: Routledge & Kegan Paul, 1972; first published 1969).

——, *The Structured Crowd: Essays in English Social History* (Brighton: Harvester Press, 1981).

Piggott, Stuart, *William Stukeley: An Eighteenth Century Antiquary*, rev. edn (London: Thames & Hudson, 1985.

Plomer, H.R., 'English Almanacs and Almanac Makers of the Seventeenth Century', *Notes and Queries* 11 (1885) 221–2, 262–4, 301–2, 382–4.

——, 'A Printer's Bill in the Seventeenth Century', *Library* 7 (1906) 32–45.

——, *A Dictionary of Booksellers and Printers . . . from 1641 to 1667* (London: Bibliographical Society, 1907).

Plongeron, Bernard (Ed.), *La Réligion Populaire dans l'occident Chrétian* (Paris: Beauchesne, 1976).

Plumb, J. H., *England in the Eighteenth Century* (Harmondsworth: Penguin, 1963).

——, 'The Commercialization of Leisure', pp. 265–85 in McKendrick et al. (1982).

Pocock, J. G. A, *The Ancient Constitution and Feudal Law: A Study in English Historical Thought in the Seventeenth Century* (Cambridge: Cambridge University Press, 1957).

——, *The Machievellian Moment: Florentine Political Thought and the Atlantic Republican Tradition* (Princeton: Princeton University Press, 1975).

——, 'Authority and Property: The Question of Liberal Origins', pp. 331–54 in Malament (1980).

——, 'No Room for the Righteous', *Times Literary Supplement* (28 December–3 January 1984) 1494.

Pollard, A. W. and Redgrave, G. R. (Eds), *A Short-Title Catalogue of Books Printed . . . 1475–1640* (London: Bibliographical Society, 1926; 2 vols).

Pollard, G., 'The English Market for Printed Books,' *Publishing History* 4 (1978) 7–48.

Popkin, Richard H., 'Predicting, Prophecying, Divining, and Foretelling from Nostradamus to Hume', *History of European Ideas* 5 (1984) 117–35.

Porter, Roy, 'Science, Provincial Culture and Public Opinion in Enlightenment England', *British Journal for Eighteenth Century Studies* 3 (1980) 20–46.

——, *English Society in the Eighteenth Century* (Harmondsworth: Penguin, 1982).

——, '"Under the Influence": Mesmerism in England', *History Today* 35 (1985) 22–29.

——, 'Lay Medical Knowledge in the Eighteenth Century: The Evidence of the Gentleman's Magazine', *Medical History* 29 (1985a) 138–68.

——, 'The Secrets of Generation Display'd: *Aristotle's Masterpiece* in Eighteenth Century England', pp. 1–21 in Maccubbin (Ed.) (1985b).

Porter, Roy and Teich, Mikulas (Eds), *Revolution in History* (Cambridge: Cambridge University Press, 1987).

Powell, Anthony, *John Aubrey and his Friends* (London: Eyre & Spottiswoode, 1948).

Poynter, F. N. L., 'Nicholas Culpeper and his Books', *Journal of the History of Medicine* 17 (1962) 152–67.

——, 'Nicholas Culpeper and the Paracelsians', pp. 201–20 in Debus (Ed.) (1972).

Prior, M.E., 'Joseph Glanvill, Witchcraft and Seventeenth Century Society', *Modern Philology* 30 (1932) 167–93.

Proust, Marcel, *Remembrance of Things Past* (Harmondsworth: Penguin, 1981; 3 vols).

Pumphrey, Steven, 'William Gilbert's Magnetical Philosophy, 1580–1684: The Creation and Dissolution of a Discipline', University of London Ph.D. (1986).

Purver, Margery, *The Royal Society: Concept and Creation* (London: Routledge & Kegan Paul, 1967).

Rabb, T. K., *The Struggle for Stability in Early Modern England* (Oxford: Oxford University Press, 1975).

Rabb, T. K., and Rotberg, R. I., *The New History: The 1980s and Beyond* (Princeton: Princeton University Press, 1980).

Rattansi, P. M., 'Paracelsus and the Puritan Revolution', *Ambix* 11 (1963) 24–32.

——, 'The Helmontian-Galenist Controversy in Restoration England', *Ambix* 12 (1964) 1–23.

——, 'The Intellectual Origins of the Royal Society', *Notes and Records of the Royal Society of London* 23 (1968) 129–43.

——, 'The Social Interpretation of Science in the Seventeenth Century', pp. 1–32 in Mathias (Ed.) (1972[a]).

——, 'Newton's Alchemical Studies', pp. 167–82 in Debus (Ed.) (1972[b]).

——, 'Some Evaluations of Reason and Sixteenth and Seventeenth Century Natural Philosophy', pp. 148–66 in Teich and Young (Eds) (1973).

——, 'Science and Religion in the Seventeenth Century', pp. 79–87 in Crosland (Ed.) (1975).

——, 'The Scientific Background', pp. 197–240 in Patrides and Waddington (Eds) (1980).

Rattansi, P. M., and McGuire, J. E., 'Newton and the "Pipes of Pan"', *Notes and Records of the Royal Society of London* 21 (1966) 108–43.

Reay, Barry (Ed.), *Popular Culture in Seventeenth Century England* (London: Croom Helm, 1985).

Redwood, John, *Reason, Ridicule and Religion, The Age of Enlightenment in England, 1660–1750* (London: Thames & Hudson, 1976).

Rivers, Isabel (Ed.), *Books and their Readers in Eighteenth Century England* (Leicester: Leicester University Press, 1982).

Roberts, William, *The Earlier History of English Bookselling* (Detroit: Gale Research, 1967; first published 1889).

Ronan, Colin A., *The Ages of Science* (London: Harrap, 1966).

——, *Edmond Halley: Genius in Eclipse* (London: MacDonald, 1969).

——, *The Cambridge Illustrated History of the World's Science* (Cambridge: Cambridge University Press, 1983).

Rorty, Richard, *Philosophy and the Mirror of Nature* (Oxford: Basil Blackwell, 1980).

——, 'The Contingency of Language', ' The Contingency of Selfhood', and 'The Contingency of Community', *London Review of Books* 8 (1986) Nos 7, 8, and 13.

Rosen, Edward, 'Forms of Irrationality in the Eighteenth Century', pp. 255–88 in Pagliaro (Ed.) (1972).

——, 'Kepler's Attitude toward Astrology and Mysticism', pp. 253–72 in Vickers (Ed.) (1984).

Ross, George MacDonald, 'Occultism and Philosophy in the Seventeenth Century', pp. 95–115 in Holland (Ed.) (1985).

Rossi, Paulo, 'Hermeticism, Rationality and the Scientific Revolution', pp. 247–74 in Bonelli and Shea (Eds) (1975).

Rousseau, André, 'Sur la "Réligion Populaire": Une Perspective Sociologique', pp. 255–60 in Duboscq et al. (Eds) (1980).

Rousseau, G. S., 'Scientific Books and their Readers in the Eighteenth Century', pp. 197–255 in Rivers (Ed.) (1982).

Rousseau, G. S. and Porter, Roy (Eds), *The Ferment of Knowledge: Studies in the Historiography of Eighteenth Century Science* (Cambridge: Cambridge University Press, 1980).

Rowse, A. L., *Simon Forman: Sex and Society in Shakespeare's Age* (London: Weidenfeld & Nicolson, 1974).

Royle, Edward, and Walvin, James, *English Radicals and Reformers 1760–1848* (Brighton: Harvester Press, 1982).

Rule, John, *The Labouring Classes in Early Industrial England, 1750–1850* (London: Longman, 1986).

Rusche, Harry, 'Merlini Anglici: Astrology and Propaganda from 1644 to 1651', *English Historical Review* 80 (1965) 322–33.

——, 'Prophecies and Propaganda,' *English Historical Review* 84 (1969) 752–70.

Russell, Colin, *Science and Social Change 1700–1900* (London: Macmillan, 1983).

Said, Edward, *The World, the Text, and the Critic* (London: Faber, 1984).

Saintyves, Pierre [Nourry, Emile], *L'Astrologie Populaire: Estudiée Spécialement dans les Doctrines et les Traditions Relatives à l'Influence de la Lune* (Paris: Emile Nourry, 1937).

Samuel, Raphael (Ed.), *People's History and Socialist Theory* (London: Routledge & Kegan Paul, 1981); including his 'History and Theory' (pp. xi–1).

——, 'Mr. Benn Consults some Household Gods', *Guardian* (4 October 1984). Samuel, Raphael and Jones, Gareth Stedman (Eds), *Culture, Ideology and Politics* (London: Routledge & Kegan Paul, 1982).

Sarton, George, *A History of Science: Ancient Science through the Golden Age of Greece* (London: Oxford University Press, 1952).

Saxl, Fritz, *Lectures*, Vol. I (London: Warburg Institute, 1957).

Schaffer, Simon, 'Natural Philosophy', pp. 55–92 in Rousseau and Porter (Eds) (1980).

——, 'Natural Philosophy and Public Spectacle in the Eighteenth Century', *History of Science* 21 (1983) 1–43.

——, 'Making Certain', *Social Studies of Science* 14 (1984) 137–52.

——, 'Occultism and Reason', pp. 117–43 in Holland (Ed.) (1985).

——, 'Authorized Prophets: Comets and Astronomers After 1759', *Studies in Eighteenth Century Culture* 17 (1987) 45–74.

——, 'Newton's Comets and the Transformation of Astrology', pp. 219–44 in Curry (Ed.) (1987a).

——, 'Godly Men and Mechanical Philosophers: Souls and Spirits in Restoration Natural Philosophy', *Science in Context* 1:1 (1987b) 55–86.

Schmitt, Charles, 'Reappraisals in Renaissance Science', *History of Science* 16 (1978) 200–14.

Schofield, Robert E., *Mechanism and Materialism: British Natural Philosophy in an Age of Reason* (Princeton: Princeton University Press, 1970).

Schott, Heinz Z. (Ed.), *Franz Anton Mesmer und die Geschichte des Mesmerismus* (Stuttgart: Franz Steiner, 1985).

Schuchard, Marsh Keith, 'Freemasons, Secret Societies, and the Continuity of the Occult Traditions in English Literature', University of Texas at Austin Ph.D. (1975).

Schwartz, Hillel, *The French Prophets: The History of a Millenarian Group in Eighteenth Century England* (Berkeley, CA: University of California, 1980).

Scienza, Credenze Occulte, Livelli di Cultura (Firenze: Istituto Nazionale di Studi sul Rinascimento, 1982).

Scribner, Bob, 'Religion, Society and Culture: Reorienting the Reformation', *History Workshop* 143 (1982) 2–22.

Secord, James A., 'Newton in the Nursery: Tom Telescope and the Philosophy of Tops and Balls, 1761–1838', *History of Science* 23 (1985) 127–51.

Sennett, Richard, *The Fall of Public Man*, (Cambridge: Cambridge University Press, 1974).

Seymour, Percy, *Astrology: The Evidence of Science* (London: Lennard, 1988).

Seznec, Jean, *The Survival of the Pagan Gods: The Mythological Tradition and its Place in Renaissance Humanism and Art* (Princeton, NJ: Princeton University Press, 1953).

Shapin, Steven, 'Social Uses of Science', pp. 93–139 in Rousseau and Porter (Eds) (1980).

——, 'History of Science and its Sociological Reconstructions', *History of Science* 20 (1982) 157–211.

Shapin, Steven and Schaffer, Simon, *Leviathan and the Air-Pump: Hobbes, Boyle and the Experimental Life* (Princeton: Princeton University Press, 1985).

Shapiro, Barbara, *Probability and Certainty in Seventeenth Century England* (Princeton, NJ: Princeton University Press, 1983).

Sharpe, J. A., *Early Modern England: A Social History, 1550–1760* (London: Edward Arnold, 1987).

Sheils, W.J. (Ed.), *The Church and Healing* (Oxford: Basil Blackwell, 1982).

Sheynin, O.B., 'On the Prehistory of the Theory of Probability', *Archive for History of Exact Sciences* 12 (1974) 99–141.

Shorr, Philip, *Science and Superstition in the Eighteenth Century. A Study of the Treatment of Science in Two Encyclopedias of 1725–1750* (New York: Columbia University Press, 1932).

Shumaker, Wayne, *The Occult Sciences in the Renaissance: A Study in Intellectual Patterns* (Berkeley, CA: University of California, 1972).

Simerly, Carol Inger, 'Origins of Astrological Terminology in English', *Astrological Journal 24 (Spring 1982) 93–9*.

Simon, Gérard, 'Kepler's Astrology: The Direction of a Reform', pp. 439–48 in Beer and Beer (Eds) (1975).

——, *Kepler: astronome, astrologue* (Paris: Gallimard, 1979).

Simon, Roger, *Gramsci's Political Thought: An Introduction* (London: Lawwrence and Wishart, 1982).

Singleton, Charles S. (Ed.), *Art, Science and History in the Renaissance* (Baltimore, MD: Johns Hopkins University Press, 1967).

Skinner, Q. R. D., 'Meaning and Understanding in the History of Ideas', *History and Theory* 8 (1969) 3–53.

—— (Ed.), *The Return of Grand Theory in the Human Sciences* (Cambridge: Cambridge University Press, 1985).

Snell, K. D. M., *Annals of the Labouring Poor. Social Change and Agrarian England, 1660–1900* (Cambridge: Cambridge University Press, 1985).

Snyder, Henry L., 'The Circulation of Newspapers in the Reign of Queen Anne', *Library* 23 (1968) 206–35.

Spierenburg, Pieter, *The Spectacle of Suffering: Executions and the Evolution of Repression* . . . (Cambridge: Cambridge University Press, 1984).

Stafford, William (Ed.), *Socialism, Radicalism and Nostalgia: Social Criticism in Britain 1775–1830* (Cambridge: Cambridge University Press, 1987).

Stahlman, William D., 'Astrology in Colonial America: An Extended Query', *William and Mary Quarterly* 13 (1956) 551–63.

Stephen, Sir Leslie and Lee, Sir Sidney (Eds), *Dictionary of National Biography* (Oxford: Oxford University Press, 1922; 22 vols).

Stevenson, John, 'The "Moral Economy" of the English Crowd: Myth and Reality', pp. 218–38 in Fletcher and Stevenson (Eds) (1981).

Stewart, Larry, 'Samuel Clarke, Newtonianism, and the Factions of post-Revolutionary England', *Journal of the History of Ideas* 42 (1981) 53–72.

——, 'Public Lectures and Private Patronage in Newtonian England', *Isis* 77 (1986) 47–58.

Stone, Lawrence, 'The Residential Development of the West End of London in the Seventeenth Century', pp. 167–212 in Malament (Ed.) (1980).

Stubbs, Mayling, 'John Beale, Philosophical Gardener of Herefordshire, Part I: Prelude to the Royal Society (1608–1663)', *Annals of Science* 39 (1982) 463–89.

Sutherland, James R., 'The Circulation of Newspapers and Literary Periodicals', *Library* 15 (1935) 110–24.

Tallmadge, G. Kasten, 'On the Influence of the Stars on Human Birth', *Bulletin of the History of Medicine* 13 (1943) 251–67.

Taylor, E. G. R, *The Mathematical Practitioners of Tudor and Stuart England* (Cambridge: Cambridge University Press, 1954).

——, *The Mathematical Practitioners of Hanoverian England, 1714–1840* (Cambridge: Cambridge University Press, 1966).

Taylor, F. Sherwood, 'An Alchemical Work of Sir Isaac Newton', *Ambix* 5 (1956) 60–84.

Teague, B. C., 'The Origins of Robert Boyle's Philosophy', Cambridge University Ph.D. (1971).

Teich, Mikulas and Young, Robert (Eds), *Changing Perspectives in the History of Science* (London: Heinemann, 1973).

Tester, Jim, *A History of Western Astrology* (Woodbridge: Bondell Press, 1987).

Thirsk, Joan, *The Rural Economy of England: Collected Essays* (London: Hambledon, 1984).

Thomas, Keith, *Religion and the Decline of Magic* (Harmondsworth: Penguin, 1973; first published 1971).

——, 'An Anthropology of Religion and Magic, II', *Journal of Interdisciplinary History* 6 (1975) 91–109.

——, *Man and the Natural World* (London: Allen Lane, 1983).

Thompson, C. J. S., *The Quacks of Old London* (London: Brentano's, 1928).

Thompson, E. P., 'Time, Work-Discipline and Industrial Capitalism', *Past and Present* 38 (1967) 56–97.

——, *The Making of the English Working Class* (Harmondsworth: Penguin, 1968; first published 1963).

——, 'Anthropology and the Discipline of Historical Context', *Midland History* 1 (1972) 41–55.

——, 'Patrician Society, Plebeian Culture', *Journal of Social History* 7 (1974) 382–405.

——, *Whigs and Hunters: The Origin of the Black Act* (Harmondsworth: Penguin, 1975).

——, 'Folklore, Anthropology and Social History', *Indian Historical Review* 3 (1977) 247–66.

——, 'Eighteenth Century English Society: Class Struggle without Class?' *Social History* 3 (1978) 133–65.

——, *The Poverty of Theory* (London: Merlin, 1978a).

Thompson, Flora, *Lark-Rise to Candleford* (Oxford: Oxford University Press, 1945).

Thorndike, Lynn, *A History of Magic and Experimental Science* (New York: Columbia University Press, 1923–58; 8 vols).

——, 'The True Place of Astrology in the History of Science', *Isis* 46 (1955) 273–78.

Timson, W. Bro. D., 'Ebenezer Sibly. Freemason Extraordinary', *Transactions* No. 2429 (1964–5) 62–7; Lodge of Research, Leicester.

Trout, Paul A., 'Magic and Millenialism: A Study of the Millenary Motifs in the Occult Milieu of Puritan England, 1640–1660', University of British Columbia D.Phil. (1975).

Underdown, David, 'Community and Class: Theories of Local Politics in the English Revolution', pp. 147–65 in Malament (Ed.) (1980).

——, *Revel, Riot and Rebellion: Popular Politics and Culture in England, 1603–60* (Oxford: Clarendon Press, 1985).

Valenze, Deborah M., 'Prophecy and Popular Literature in Eighteenth Century England', *Journal of Ecclesiastical History* 29:1 (1978) 75–92.

Veyne, Paul, *Comment on écrit l'histoire* (Paris: de Seuil, 1971; with another edition in 1978, to which is added *Foucault révolutionne l'histoire*); translated as *Writing History. Essays on Epistemology* (Manchester: Manchester University Press, 1984).

Vickers, Brian, 'Frances Yates and the Writing of History', *Journal of Modern History* 51 (1979) 287–316.

—— (Ed.), *Occult and Scientific Mentalities* (Cambridge: Cambridge University

Press, 1984); including his 'Analogy vs. Identity: The Rejection of Occult Symbolism, 1580–1680' (pp. 95–165).

Vilar, Pierre, 'Constructing Marxist History', pp. 47–80 in Le Goff and Nora (Eds) (1985).

Vincent, David, *Literacy and Popular Culture: England 1750–1914* (Cambridge: Cambridge University Press, 1989).

Vovelle, Michel, 'Ideologies and Mentalities', pp. 2–11 in Samuel and Jones (Eds) (1982) [a translation of pp. 5–17 in Vovelle (1985)].

——, *Idéologies et Mentalités* (Paris. La Découverte, 1985; first published in 1982; to be published in English by Polity Press, 1989.

Walker, D. P., *Spiritual and Demonic Magic from Ficino to Campanella* (London: Warburg Institute, 1958).

Wallis, Roy (Ed.), *On the Margins of Science: The Social Construction of Rejected Knowledge* (Keele: University of Keele, 1979).

Ward, Eric, 'Ebenezer Sibly – A Man of Parts', *Ars Quator Coronatorum* 71 (1958) 48–52.

Watts, Sheldon, *A Social History of Western Europe 1450–1720: Tensions and Solidarities among Rural People* (London: Hutchinson, 1985).

Weaver, Helen, with Brau, Jean-Louis, and Edmands, Allan (Eds), *Larousse Encyclopedia of Astrology* (New York: Larousse, 1980; first published in Paris in 1977).

Webb, Alfred, *A Compendium of Irish Bibliography* (Dublin: M. H. Gill, 1878).

Webster, Charles, 'The Origins of the Royal Society', *History of Science 6* (1967) 106–28.

——, *Samuel Hartlib and the Advancement of Learning* (Cambridge: Cambridge University Press, 1970).

——, *The Great Instauration: Science, Medicine and Reform, 1626–1660* (London: Duckworth, 1975).

——, 'Paracelsus and Demons: Science as a Synthesis of Popular Belief', pp. 3–20 in *Scienza, Credenze Occulte, Livelli di Cultura* (1982).

——, *From Paracelsus to Newton: Magic and the Making of Modern Science* (Cambridge: Cambridge University Press, 1982a).

Wedel, Theodore Otto, *The Mediaeval Attitude Toward Astrology: Particularly in England* (New Haven, CT: Yale University Press, 1920).

Westfall, Richard, 'Newton and the Hermetic Tradition', pp. 183–98 in Debus (Ed.) (1972) vol. II.

——, 'The Role of Alchemy in Newton's Career', pp. 189–232 in Bonelli and Shea (Eds) (1975).

——, *Never at Rest: A Biography of Isaac Newton* (Cambridge: Cambridge University Press, 1980).

——, 'Newton and Alchemy', pp. 315–36 in Vickers (Ed.) (1984).

Westman, Robert S. and McGuire, J. E., *Hermeticism and the Scientific Revolution* (Los Angeles: University of California, 1977).

Whiteside, D. T., 'Before *The Principia*: The maturing of Newton's Thoughts on Dynamical Astronomy', *Journal for the History of Astronomy* 1 (1970) 5–19.

——, 'Newton the Mathematician', pp. 109–27 in Bechler (Ed.) (1982).

Wilde, C. B., 'Hutchinsonianism, Natural Philosophy and Religious Controversy in Eighteenth Century Britain', *History of Science* 18 (1980) 1–24.

——, 'Matter and Spirit as Natural Symbols in Eighteenth Century British Natural Philosophy', *British Journal for the History of Science* 15 (1982) 99–131.

Williams, John Peregrine, 'The Making of Victorian Psychical Research: An Intellectual Elite's Approach to the Spiritual World', Cambridge University D.Phil. (1984).

Williams, Raymond, *Keywords* (London: Fontana, 1976).

——, *Marxism and Literature* (Oxford: Oxford University Press, 1977).

——, *Politics and Letters: Interviews with the New Left Review* (London: New Left Books, 1979).

Wilson, Adrian, 'The Infancy of the History of Childhood: An Appraisal of Philippe Ariès', *History and Theory* 19 (1980) 132–53.

Wilson, Adrian and Ashplant, T. G., 'Whig History and Present-Centred History', *Historical Journal* 31:1 (1988) 1–16.

Wing, Donald (Ed.), *Short-Title Catalogue of Books Printed . . . 1641–1700* (New York: Columbia University Press, 1945–51; 3 vols).

Wirth, Jean, 'Against the Acculturation Thesis', pp. 66–78 in Greyerz (Ed.) (1984).

Wood, Chauncey, *Chaucer and the Country of the Stars: Poetic Uses of Astrological Imagery* (Princeton, NJ: Princeton University Press, 1970).

Wood, Paul, 'Francis Bacon and the Experimental Philosophy: A Study in Seventeenth Century Methodology', University of London M.Phil. (1978).

Woolf, Harry (Ed.), *The Analytic Spirit: Essays in the History of Science* (Ithaca, New York: Cornell University Press, 1981).

Wright, Peter W. G., 'Astrology and Science in Seventeenth Century England', *Social Studies of Science* 5 (1975) 399–422.

——, 'A Study in the Legitimization of Knowledge: The "Success" of Medicine and the "Failure" of Astrology', pp. 85–102 in Wallis (Ed.) (1979).

——, 'Astrology in Mid-Seventeenth Century England: A Sociological Analysis', University of London Ph.D. (1984).

Wrightson, Keith, *English Society 1580–1680* (London: Hutchinson, 1982).

——, 'The Social order of Early Modern England: Three Approaches', pp. 177–202 in Bonfield et al. (1986).

Wrigley, E. A., 'A Simple Model of London's Importance in Changing English Society and Economy 1650–1750', *Past and Present* 37 (1967) 44–70.

Wynne, Brian, 'Physics and Psychics: Science, Symbolic Action, and Social Control in Late Victorian England', pp. 167–86 in Barnes and Shapin (1979).

Yates, Frances, *Giordano Bruno and the Hermetic Tradition* (London: Routledge & Kegan Paul, 1964).

——, 'The Hermetic Tradition in Renaissance Science', pp. 255–74 in Singleton (Ed.) (1967).

——, *The Rosicrucian Enlightenment* (London: Routledge & Kegan Paul, 1972).

Yeo, Stephen and Eileen, *Popular Culture and Class Conflict, 1590–1914 . . .* (Brighton: Harvester Press, 1981).

Yolton, John W., *Thinking Matter: Materialism in Eighteenth Century Britain* (Oxford: Basil Blackwell, 1984).

Zambelli, Paola, 'Uno, due, tre, mille Menochio?' *Archivo Storico* 1 (1979) 51–90.

——, 'Fine del Mondo o Inizio Della Propaganda? Astrologia, filosofia della storia e propaganda politico-religiosa nel di battito sulla congiunziomne del 1524', pp. 291–368 in *Scienza, Credenze Occulte, Livelli di Cultura* (1982).

Zoller, Robert, 'Aristotelianism and Hermeticism in Medieval Astrology', *Geocosmic Research Monograph* No. 3 (1982) 24–28.

Index

Index

medicine (physic) 21, 23–5, 41–2, 55–6, 79, 99–101, 109, 151

mentality (*mentalité*) 3, 104–5, 113–15, 158–62, 186 (n. 26)

Merrifield, John 88

Mesmerism 116, 131, 151

Methodism 112, 164

Moon 11, 97–9, 121

Moore's Almanack (Vox Stellarum) 56, 101–2, 113–17, 120, 125, 137, 158, 161–2

Moore, Francis 25, 56, 101

More, Henry 49–50, 54

Morin, Jean-Baptiste 86–7

Moxon, Joseph 43

natural philosophy *see* science

Newton, Isaac (Newtonianism) 34, 40, 131, 139, 142–7

Obelkevitch, James 3–4, 169 (n. 4)

Oldenburg, Henry 66, 74

Overton, Richard 26

Parker, George 38, 76–7, 81, 160

Parker, Samuel 49–50, 160

Partridge, John 31, 44, 54, 57, 73, 79–86, 89–91, 136, 160

Platonism (neo-) 9, 49, 104

Ptolemy 10, 12–13, 15, 77, 81–2, 84–5, 95, 120

Purslow, Norris 104

rationality 4, 14–15

Ray, John 145

Reeve, Edmund 42

relativism 4, 189 (n. 55)

Restoration 6–7, 39, 43, 45–6, 48, 52, 58, 77–8

Revolution, English (Interregnum) 2, 6, 19–20, 45–6, 58, 155–6

Revolution, Glorious 2, 6, 54, 106

Rollison, Oud 103

Royal Society 6, 27, 32, 39–40, 48, 50, 54, 57–60, 64, 67–8, 76–8, 110–12, 139

Salmon, William 21, 25, 56, 79, 99–100

Saunders (Sanders), Richard 21, 24–5, 27, 41–3, 55

Schaffer, Simon 58, 142, 184 (n. 10)

science (natural philosophy) 31–4, 39–40, 57–8, 136–52, 166

Season, Henry 96, 119, 126–7

sects/sectarians 25–7, 29, 45–7, 141

Shakerley, Jeremy 32–3

Sharp, Edward 122

Sibly, Ebenezer 134–7

Southcott, Joanna 116

Southey, Robert 98–9, 103

Spectator 106, 110–11

Sprat, Thomas 60, 160

The Stranger in Reading 102

Streete, Thomas 34–5, 53, 66

Stukeley, William 122–4

Swadlin, Thomas 42

Swift, Jonathan 80, 89–91, 106, 110–11, 160, 164

Tatler 106, 110–11, 160

Thomas, Keith 2, 7, 29

Thompson, E. P. 2, 117, 154–6, 185 (n. 13)

Thompson, Flora 97

Thrale, Mrs Hester 95, 128

Titus, Placidus 13, 83

Turner, Richard 150

Veyne, Paul 156

Vovelle, Michel 159

Walton, Richard 104–5

Ward, Seth 27, 59, 64

237